COMPUTING BRAIN ACTIVITY MAPS FROM fMRI TIME-SERIES IMAGES

fMRI is a very popular method in which researchers and clinicians can image human brain activity in response to given mental tasks. This book presents a comprehensive review of the methods for computing activity maps, while providing an intuitive and mathematical outline of how each method works. The approaches include statistical parametric maps (SPM), hemodynamic response modeling and deconvolution, Bayesian, Fourier and nonparametric methods. The newest activity maps provide information on regional connectivity and include principal and independent component analysis, crisp and fuzzy clustering, structural equation modeling, and dynamic causal modeling. Preprocessing and experimental design issues are discussed with references made to the software available for implementing the various methods. Aimed at graduate students and researchers, it will appeal to anyone with an interest in fMRI and who is looking to expand their perspectives of this technique.

GORDON E. SARTY is Associate Professor in the Department of Psychology at the University of Saskatchewan.

COMPUTING BRAIN ACTIVITY MAPS FROM fMRI TIME-SERIES IMAGES

GORDON E. SARTY

University of Saskatchewan

CAMBRIDGE
UNIVERSITY PRESS

CAMBRIDGE UNIVERSITY PRESS
Cambridge, New York, Melbourne, Madrid, Cape Town, Singapore, São Paulo

Cambridge University Press
The Edinburgh Building, Cambridge CB2 2RU, UK

Published in the United States of America by Cambridge University Press, New York

www.cambridge.org
Information on this title: www.cambridge.org/9780521868266

First published 2007

Printed in the United Kingdom at the University Press, Cambridge

A catalog record for this publication is available from the British Library

ISBN-13 978-0-521-86826-6 hardback
ISBN-10 0-521-86826-2 hardback

To Kerry

Contents

Contents

The color plates are situated between pages 38 and 39

Preface

You will find here a comprehensive review of all fMRI data processing methods proposed in the literature to 2005. I have endeavored, however, to produce a review that is useful to more than those already in the know. With the introduction of each major method I have additionally given an overview of how the method works in wide and hopefully intuitive terms. The overviews taken together should give the newcomer a broad idea of all the choices that can be made for transforming raw fMRI data into a brain activity map.

The term activity map is used here to include the specific forms of maps labeled as activation maps and connectivity maps. Activation maps show regions of the brain that are active in response to a task given to the person in the MRI, while connectivity maps are intended to provide information on the neural connectivity between the active regions.

All methods are described in a precise manner from a mathematical perspective. So a certain amount of mathematical and/or statistical background is assumed of the reader. However, the math is presented at a high level. You will not find here details of how to implement any of the methods reviewed. For the details you will need to consult the original literature listed in the references.

In short, this book can help the newcomer to the field of fMRI (or a seasoned researcher wanting to know about methods used by others) to become oriented via a three-step process. The first step is to more or less skim through the book, reading the descriptive overviews to get a perspective on the big picture. Then, after some methods of interest are identified, a more careful reading of the high level mathematical descriptions can be done to become more familiar with the underlying ideas. Finally, if the reader is convinced of the utility of a particular method, the original literature can be consulted, the methods mastered and contributions of one's own to the field envisaged. This is the path that all graduate students using fMRI are interested in taking so I imagine that it will be those graduate students who will find this book most useful.

This book (in a slightly shorter form) was originally written as a review article for *Current Medical Imaging Reviews* (*CMIR*) and submitted for peer review. The finished manuscript proved to be longer than I originally envisaged. I was under the impression when I first began to write that there were only a "few" basic ways to process fMRI data. That impression was proved to be very wrong as I dug into the literature to find literally hundreds of ways of turning raw fMRI data into something interpretable. So, while the reviews of the original manuscript came back from *CMIR* generally positive, it was far too long to be published in the journal. At that point the editor for *CMIR*, Sohail Ahmed, agreed to support the publication of the review in book form by Cambridge University Press. With thanks to Martin Griffiths at Cambridge University Press this book has subsequently become a reality. Thanks are due also to Clare Georgy for her help in the details of producing the final manuscript and Betty Fulford for looking after the legal aspects. Thanks to my technical assistant, Jennifer Hadley, here at the University of Saskatchewan, for preparing some of the figures and for organizing copies of all the references into an ordered stack of paper that stood at a height just slightly less than my own. Thanks, finally and mostly, to my family: Kerry, Dominic and Darien, for their patience and understanding with the long night hours I devoted to writing this review.

1
Introduction

The production of a brain activity map from data acquired with a volunteer or patient and magnetic resonance imaging (MRI) [269] requires a fairly wide range of interdisciplinary knowledge and techniques. Producing brain activity maps from functional MRI (fMRI) data requires knowledge and techniques from cognitive neuropsychology, physics, engineering and mathematics, particularly statistics. The process typically begins with a question in cognitive psychology that can be answered, at least in part, by combining the knowledge obtained from brain activity maps with previous knowledge of the function of specific regions of the brain. The previous knowledge of regional brain function generally has its origin in lesion studies where disease or injury has removed a brain region, and its function, from the brain's owner. Such lesion-based knowledge has firmly established the principle of *functional segregation* in the brain, where specific regions are responsible for specific functions. The use of fMRI to produce *activation maps* allows specific questions on functional segregation to be posed and investigated without risk to the person being studied. The brain is also known to be a very complex system in which several regions, working cooperatively, are required for some tasks. This cooperation among regions is known as *functional integration* and may be studied using fMRI techniques that lead to *connectivity maps*. Methods for producing activation and connectivity maps are reviewed here with the goal of providing a complete overview of all data processing currently available to produce the brain activity maps from raw fMRI data. This overview should be of particular use to those wishing to begin or revise a program of fMRI investigation, whether from a clinical perspective or from a research perspective, by setting out and explaining the various data processing options now available.

A team of fMRI investigators generally includes an expert in the field of cognitive science and an engineering/physics expert (rarely the same person). The overview will primarily, but not exclusively, be useful for the engineer/physicist on the team. The overview will also be of value to those who wish to contribute to fMRI data analysis methodology by providing a summary of what has been accomplished to date by others.

1.1 Overview of fMRI technique

With a sufficiently sharp question about cognition, the experimental design begins with the design of a "task paradigm" in which specific sensory stimuli are presented to the volunteer in the MRI apparatus. The stimuli are frequently visual (e.g. images or words) or auditory (speech) in nature and are presented using computer hardware and software designed specifically for cognitive science experimentation. The software E-Prime (Psychology Software Tools, Inc., Pittsburgh, USA) is commonly used for such presentations although the job can be done fairly easily from scratch using something like Visual Basic. The most important aspect of the presentation software, aside from submillisecond timing accuracy, is its ability to either send triggers to or receive them from the MRI[†] to keep the presentation in synchrony with the image acquisition[‡].

While the volunteer is performing the task, the MRI apparatus is gathering image data of the volunteer's brain. To acquire image data a *pulse sequence* program is required to run on the MRI's computer. The pulse sequence commands the switching of the MRI's magnetic gradient coils[§], radio frequency (RF) transmission and reception and the acquisition of data to microsecond timing accuracy. We will focus here on echo planar imaging (EPI) [297] pulse sequences that acquire a time-series of volume (3D) image data typically at intervals of 1–2 s/volume (this value can vary from experiment to experiment and is known as the repeat time (T_R) of the imaging experiment). Non-EPI approaches to functional brain imaging are reviewed in [251]. We include in the class of EPI sequences the traditional Cartesian approach plus those that depend on other trajectories in k-space[¶] like spirals [6], rosettes [344, 386], Lissajous patterns [323] or radial patterns [388, 400], because the resulting data sets are essentially the same from a brain activity map data processing point of view.

[†] We will also use MRI to designate magnetic resonance imager.

[‡] Most investigators prefer that the trigger comes from the MRI because the timing from the MRI is usually more accurate than timing from software running on a typical microcomputer – especially microcomputers that devote significant resources to maintenance of the operating system.

[§] The banging or pinging noise heard during the operation of an MRI is from switching large amounts of current ($\sim 10^2$ A) in the gradient coils. From an engineering point of view, it is possible to build gradient coils that do their job silently [89, 90, 129] but there has not yet been sufficient economic motivation for MRI manufacturers to supply a silent MRI apparatus.

[¶] MRI data are acquired in a spatial frequency space known as k-space and must be reconstructed into an image using Fourier transformation techniques [385]. The image reconstruction process is usually invisible to the fMRI investigator, having been done by software installed in the MRI.

A volume data set is made up of a collection of volume elements or *voxels*, a term generalized from the word pixel (picture elements) used in electronic imaging. An fMRI data set consists of a time-series of volume data sets, so each data point in an fMRI data set has a unique coordinate (x, y, z, t) in 4D space, $\mathbb{R}^4$, where x, y and z are spatial coordinates and t is time. We will also focus exclusively on functional imaging techniques that rely on the blood oxygenation level dependent (BOLD) phenomenon [347]. BOLD works because deoxygenated hemoglobin is paramagnetic while oxygenated hemoglobin is not paramagnetic [355]. The presence of paramagnetic deoxyhemoglobin causes the signal-producing proton spins to dephase more rapidly than they would otherwise because of the local magnetic field gradients caused by the paramagnetism. The dephasing mechanism is a partly reversible process known as T_2^* if a gradient echo EPI sequence is used, or an irreversible process known as T_2 if a spin echo EPI sequence is used. An increase in blood flow to active brain regions results in a higher concentration of oxygenated blood and therefore results in an increased MRI signal in the active region [348].

1.2 Overview of fMRI time-series data processing

So the stage is set for the mathematical analysis of our fMRI data set: we have in our possession a time-series of EPI images in which we hope to see evidence of the BOLD effect in voxels of interest (Fig. 1.1). Before analysis, the data set may require some preprocessing to remove the effects of motion, "noise" and intersubject variation of neuroanatomy. Relevant image preprocessing approaches are reviewed in Chapter 2. Next, before the data can be analyzed to produce activation maps, we need to understand the task paradigm. There are two basic task paradigm designs, blocked and event-related; these designs are reviewed in Chapter 3.

By far the most common approach to activation map computation involves a univariate analysis of the time-series associated with each (3D) voxel. The majority of the univariate analysis techniques can be described within the framework of the general linear model (GLM) as we review in Section 4.1. Many GLMs depend on a model of the hemodynamic response which may, in turn, be modeled mathematically. Models of the hemodynamic response and their application to fMRI data are reviewed in Section 4.2. Other methods for producing activation maps include various other parametric methods (Section 4.3), Bayesian methods (Section 4.4), nonparametric methods (Section 4.5) and Fourier methods (Section 4.6). Of particular interest to clinical investigators, who will want to make medical decisions on the basis of an fMRI investigation, is the reliability of the computed activation maps. Some of these reliability issues, from the point of view of repeatability, are covered in Section 4.7. Finally, in a clinical setting, speed is important for many reasons. Additionally, fMRI may be used along with other approaches to gather

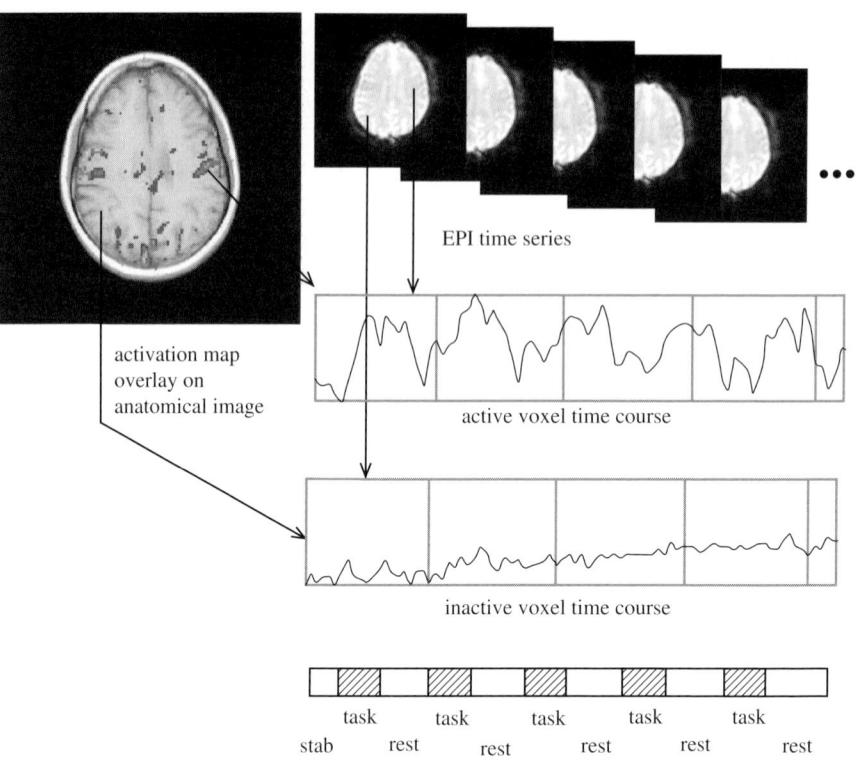

EPI time series

activation map
overlay on
anatomical image

active voxel time course

inactive voxel time course

stab   task   rest   task   rest   task   rest   task   rest   task   rest

Fig. 1.1. Schematic of a typical EPI BOLD fMRI data set and its reduction to an activation map. An EPI time-series, represented here by a series of single slices from the volume data set at each point in time, is collected while the subject in the MRI performs a task synchronized to the image collection. The intensity of each voxel in the image set will vary in time. An active voxel will vary in step with the presented task while an inactive voxel will not. A statistical method is needed to determine if a given voxel time course is related to the task, and therefore active, or not. The active voxels are then color coded under a relevant amplitude–color relationship and the resulting activation map is overlain on a high resolution anatomical MRI image to allow the investigator to determine which brain regions are active under the given task. See also color plate.

converging evidence. Real time aspects and the use of fMRI with other approaches like electroencephalograms (EEGs) are briefly reviewed in Section 4.8.

From a purely statistical point of view, fMRI data should be analyzed using multivariate techniques so that correlations between the time courses of the individual voxels are taken into account. These correlations are also of interest to the cognitive neuroscientist because they carry information about connectivity. Therefore multivariate GLM approaches, of which an overview is given in Section 5.1, may be used to compute connectivity maps. A major problem with multivariate approaches is that the data vector (the 3D volume data set) has a very high dimension ($\sim 10^5$) in comparison to the number of vector data points (=time points, $\sim 10^2$) making the straightforward application of multivariate statistics impossible. The solution is to reduce the dimensionality of the data by decomposing the data into components. Components may be produced using principal component analysis (PCA) or independent component analysis (ICA) and their variations as we review in Section 5.2. When used spatially, the relatively new mathematical technique of wavelet analysis may also be used to analyze the fMRI data set as a whole. Wavelet analysis is particularly attractive for fMRI analysis because confounding physiological influences like heart beat and breathing leave behind model noise that is autocorrelated. Wavelet analysis, because of its fractal properties, naturally "whitens" the noise to decorrelate it and makes statistical testing more straightforward. Wavelet approaches to brain map computation are reviewed in Section 5.3. PCA and ICA use a specific criterion, variance and statistical independence respectively, to define components or images of related voxels. Clustering approaches use a more empirical criterion, usually based on distance in data space, to produce images of related voxels. Clustering approaches are reviewed in Section 5.4. Our review ends with a look at the issues of functional connectivity (Section 5.5) and effective connectivity (Section 5.6). The issue of effective connectivity, where fMRI data are used to make inferences about how different regions of the brain affect each other causally, represents the current cutting edge of fMRI brain map computation. Models of effective connectivity will become more sophisticated, and will be capable of modeling networked neural activity, as more becomes known about the physiology of the neural hemodynamic system.

1.3 Mathematical conventions

The following mathematical conventions and assumptions will be adopted in the presentation that follows. All functions f may be complex-valued and are assumed to be square integrable, $f \in L^2(\mathbb{R}^n)$. This allows us to use the inner product of two functions f and g

$$\langle f, g \rangle = \int_{\mathbb{R}^n} f(\vec{x})\, g^*(\vec{x})\, d\vec{x}, \tag{1.1}$$

where the $*$ denotes complex conjugation, and norm

$$\|f\|^2 = \int_{\mathbb{R}^n} |f(\vec{x})|^2\, d\vec{x}. \tag{1.2}$$

The functions that we will encounter will frequently be real-valued and have compact support, so these functions belong to $L^2(\mathbb{R}^n)$ without any additional assumptions. The correlation, r, of two functions f and g is the normalized inner product of the functions minus their mean, $F = f - \bar{f}$ and $G = g - \bar{g}$, as defined by

$$r = \frac{\langle F, G \rangle}{\|F\| \, \|G\|}, \tag{1.3}$$

where $\bar{h} = \int h(\vec{x}) \, d\vec{x}/V$ with V being the hypervolume of an appropriate domain and where $\|H\| \equiv \sigma_h$, the standard deviation of h. The cross-correlation function, c, of two functions f and g is defined by

$$c(\vec{y}) = \int_{\mathbb{R}^n} f(\vec{x}) \, g^*(\vec{x} + \vec{y}) \, d\vec{x}, \tag{1.4}$$

the autocorrelation function of f by

$$a(\vec{y}) = \int_{\mathbb{R}^n} f(\vec{x}) f^*(\vec{x} + \vec{y}) \, d\vec{x} \tag{1.5}$$

and the convolution $f * g$ by

$$f * g(\vec{y}) = \int_{\mathbb{R}^n} f(\vec{x}) \, g^*(\vec{y} - \vec{x}) \, d\vec{x}. \tag{1.6}$$

When Fourier transforms are required, we assume that $f \in L^2(\mathbb{R}^n) \cap L^1(\mathbb{R}^n)$, where $L^1(\mathbb{R}^n)$ is the space of absolutely (Lebesgue) integrable functions. Then we may define the Fourier transform $\mathcal{F}$ and its inverse, $\mathcal{F}^{-1}$ as

$$\mathcal{F}f(\vec{\eta}) = \int_{\mathbb{R}^n} f(\vec{x}) \, e^{-2\pi i \vec{x} \cdot \vec{\eta}} \, d\vec{x} \quad \text{and} \quad \mathcal{F}^{-1}g(\vec{x}) = \int_{\mathbb{R}^n} g(\vec{\eta}) \, e^{2\pi i \vec{x} \cdot \vec{\eta}} \, d\vec{\eta} \tag{1.7}$$

where $i = \sqrt{-1}$. The Fourier transform of the autocorrelation function is the spectral density function.

Vectors, $\vec{x}$, will be considered to be column vectors in $\mathbb{R}^n$ with $\vec{x}^T$ representing a row vector. To make use of functional ideas, we may also consider $\vec{x} \in \ell^2$, where ℓ^2 is the vector space of square summable sequences. In either case, the inner product and norm are defined by

$$\langle \vec{x}, \vec{y} \rangle = \sum_k x_k y_k^* = \vec{x}^T \vec{y}^* \quad \text{and} \quad \|\vec{x}\|^2 = \langle \vec{x}, \vec{x} \rangle \tag{1.8}$$

(see Equations (1.1) and (1.2)), where x_k and y_k denote the components of $\vec{x}$ and $\vec{y}$ respectively. The correlation between two vectors $\vec{x}$ and $\vec{y}$ in a K-dimensional subspace is

$$r = \frac{\langle \vec{X}, \vec{Y} \rangle}{\|\vec{X}\| \, \|\vec{Y}\|}, \tag{1.9}$$

where $\vec{A} = \vec{a} - \bar{\vec{a}}$ with $\bar{\vec{a}} = \sum_{k=1}^{K} a_k/K$ and where $\|\vec{A}\| \equiv \sigma_{\vec{a}}$ the standard deviation of $\vec{a}$ (see Equation (1.3)). The quantity $\langle X, Y \rangle$ is the covariance between $\vec{X}$ and $\vec{Y}$ (or $\vec{x}$ and $\vec{y}$). The cross-correlation and convolution of two vectors in ℓ^2 may be defined analogously to Equations (1.4) and (1.6). The discrete Fourier and inverse Fourier transforms of $\vec{x} \in \mathbb{R}^n$ are given by

$$(F\vec{x})_k = \sum_{j=1}^{n} x_j \, e^{-2\pi i k j} \quad \text{and} \quad (F^{-1}\vec{x})_k = \sum_{j=1}^{n} x_j \, e^{2\pi i k j} \tag{1.10}$$

respectively, where we note that conventions on the numbering of indices may be made differently to allow a straightforward interpretation of $(F\vec{x})_k$ as frequency components.

The convolution theorem [57, 379] states that

$$\mathcal{F}(g * f) = (\mathcal{F}g)(\mathcal{F}f) \quad \text{and} \quad F(\vec{g} * \vec{f}) = (F\vec{g})(F\vec{f}). \tag{1.11}$$

I use $\vec{x}$ to represent a finite-dimensional vector and $[X]$ to represent a matrix. There is some variation in the literature as to the use of the words, scan, session, etc. For this review the following definitions are adopted:

- *Volume:* A set of 3D voxels obtained at one time point in the fMRI time-series. A volume is composed of a number of 2D slices. A slice, synonymous with image, is composed of pixels. Pixels and voxels are really the same thing but the word voxel is generally used when an analysis is considered from a 3D point of view and pixel is used when a 2D point of view is considered.
- *Scan:* Synonymous with volume.
- *Run:* A time-series of volumes continuously obtained with a time of T_R between volume acquisition. Many authors use *session* as a synonym of run as defined here. However we reserve session for the next definition.
- *Session:* A collection of runs collected from the same subject. The subject stays in the MRI for an entire session. This use of the word session conforms to its use in the software package AFNI.
- *Subject:* Synonymous with the American Psychological Association's participant. In a repeated measures design, a subject will typically experience multiple sessions.

1.4 Note on software availability

Freely available software for many of the techniques described in this review is available either directly from the authors of the original literature or on the internet. Internet addresses are given where known. An "early" review of different software packages available in 1998 is given by Gold *et al.* [189]. That review primarily lists the "features" provided by each software package, particularly in terms of what kind of analysis can be done. As a cursory survey of the present review will indicate, there is no one best method for data analysis and therefore no one best software package to use. Some side-by-side comparisons of methods have been done; those comparisons are briefly reviewed in Section 4.7. A few commercial software packages are referenced in this review and their mention should not be taken as an endorsement. Commercial software is only referenced in cases where it is directly relevant to the reviewed literature and/or where I have personal experience with the referenced software.

2
Image preprocessing

fMRI time-series contain a number of systematic sources of variability that are not due to the BOLD effect of interest. These sources of systematic variability may be removed as a preprocessing step, as outlined in this chapter, or they may be removed as a covariate at the time of activation map computation. The sources of variability include factors due to the physics of MRI, subject motion, heart beat and breathing, other physiological processes, random thermally generated noise and intersubject anatomical variability. The variability due to the physics of MRI include ghosts, geometric distortion and some signal drift which (with the exception of drift) may be removed at the pulse sequence and image reconstruction level. However, such ideal pulse sequences are not yet widely available on commercial scanners and many investigators must simply accept compromises, such as the geometric distortion of EPI images. A small amount of subject motion can be compensated for by aligning the time-series fMRI images but motion may also be dealt with at the source by using suitable restraining devices and training the subject to hold still (mock MRIs are very useful for training study subjects in this respect). It is impossible to remove the source of physiological variables so these need to be corrected at the time of data processing. Two approaches are used to remove physiologic effects from the data: one is to model heart beat and breathing, the other is to measure and remove the effect of a global signal. Again, many investigators do not account for non-BOLD physiologic variation at the expense of reduced statistical power to detect BOLD activations. Various filtering techniques also exist to remove thermally generated noise from the time-series with the trade-off of wiping out small BOLD variation in the smoothing process. Finally, to compare activation maps between subjects, it is necessary to transform or warp the data sets into a common stereotaxic space. The image processing literature is full of many methods for registering images between subjects and between imaging modalities. Here we only touch upon a small number of approaches that have been applied to fMRI data.

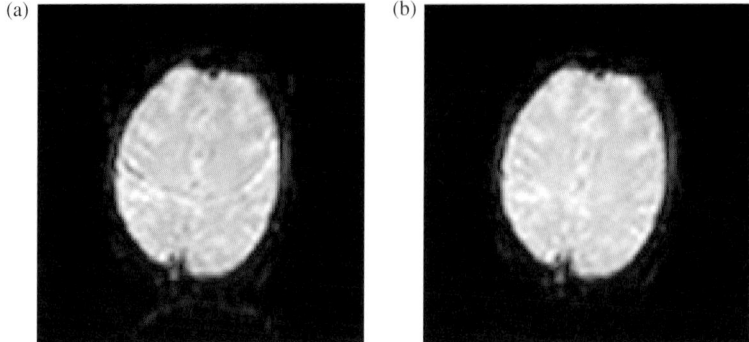

Fig. 2.1. Proton NMR signal from the scalp comes primarily from fat while signal from the brain comes primarily from water. Proton spins in fat and water emit at different resonant frequencies, because of chemical shift, and this difference is interpreted in the Fourier transform reconstruction as being due to different locations, especially in the phase encoding direction for EPI sequences. When signal from both fat and water is received, a ghost image of the scalp appears as shown in (a). Suppressing the signal from fat with an appropriately designed pulse sequence eliminates the ghost as shown in image (b) where fat saturation pulses have been used.

2.1 EPI ghost artifact reduction

Ghost images in EPI images can be caused by chemical shift artifact (primarily caused by signal from fat), misalignment of k-space lines (for conventional EPI) in the raw data set, main magnetic field inhomogeneities, or susceptibility gradients from anatomical inhomogeneity. Some of these ghost artifacts can be easily removed, while others may be next to impossible to remove so that the data will remain slightly contaminated by ghosts when they are analyzed. Many of the methods for eliminating or reducing ghost artifacts can be, and are for many commercial EPI implementations, applied at the pulse sequence level or at the Fourier transform image reconstruction stage, resulting in a raw data set for the investigator that already has corrections for ghost artifacts applied. Some methods of ghost correction require ancillary data, such as magnetic field maps, in addition to the image data, while other methods work using only the image data.

The easiest ghost artifact to reduce is that caused by the signal from fat in the scalp, see Fig. 2.1. A pulse sequence that applies a fat saturation pulse before image data acquisition will eliminate most of the chemical shift[†] ghost artifact.

The ghost artifact caused by misalignment of alternating k-space lines in conventional EPI is known as an $N/2$ ghost because the ghost image is shifted by $N/2$

[†] Chemical shift is a shifting of the emitted RF frequency from the proton due to the shielding of the main magnetic field by the electronic configuration around the proton. Chemistry depends on molecular electronic configuration hence the term chemical shift.

from the main image in the phase direction when the number of lines in the phase encode direction[†] is N. The misalignment of the odd and even k-space lines may be caused by gradient field eddy currents, imperfect pulse sequence timing, main field inhomogeneity and susceptibility gradients. Most packaged EPI sequences acquire two or three reference lines before the actual collection of image data so that the offset between the odd and even lines may be measured and a phase correction applied before Fourier image reconstruction. It is possible to determine the necessary correction from the image data directly but more computation is required. Buonocore and Gao [71] give a method for correcting the $N/2$ ghost directly from image data for single-shot conventional EPI, the type normally used for fMRI, while Buonocore and Zhu [72] give a method that works for interleaved EPI where more than one acquisition is required to obtain data for one image slice.

Non-Cartesian EPI approaches, such as spiral acquisition in k-space, do not suffer from $N/2$ ghost artifacts but instead the factors that cause $N/2$ ghosts in Cartesian EPI lead to a local blurring in non-Cartesian approaches [236].

2.2 Geometric distortion correction

The same sources that lead to ghost artifacts can also lead to geometric distortion of an EPI image. EPI is more sensitive to geometric distortion than the high resolution spin warp spin echo images that are commonly used as underlays for the activation maps. Consequently the exact source of the activations, as determined by their position on the spin echo image, may be in error by several millimeters. The sensitivity of EPI to geometric distortion stems from the relatively long acquisition time ($\sim$40 ms) required and the sensitivity to spread in the reconstruction point spread function (PSF) in the phase encode direction caused by magnetic field inhomogeneities[‡]. Modeling the phase evolution of the MRI signal and applying the reconstruction Fourier transform without any correction gives a reconstructed image ρ_1 that is related to the undistorted image ρ by

$$\rho_1(x,y) = \rho \left(x \pm \frac{\Delta B(x,y)}{G_x}, y + \frac{\Delta B(x,y)T}{G_y \tau} \right), \tag{2.1}$$

where G_x is the magnitude of the applied frequency encoding gradient, G_y is the (average) magnitude of the blip phase encoding gradient, T is the acquisition time per line of k-space and τ is the duration of the phase encoding gradient blip in the EPI pulse sequence [471]. The $\pm$ in Equation (2.1) refers to the odd and even k-space lines. Equation (2.1) shows that the geometric distortion is greatest in the

---

[†] Standard EPI images have a frequency encode direction and phase encode direction. See Vlaardingerbroek and den Boer [433] and Haacke *et al.* [200] for a complete discussion on MRI pulse sequences.

[‡] The total inhomogeneity, ΔB, at any given voxel is the sum of the effects mentioned in Section 2.1.

(a)   (b)   (c)

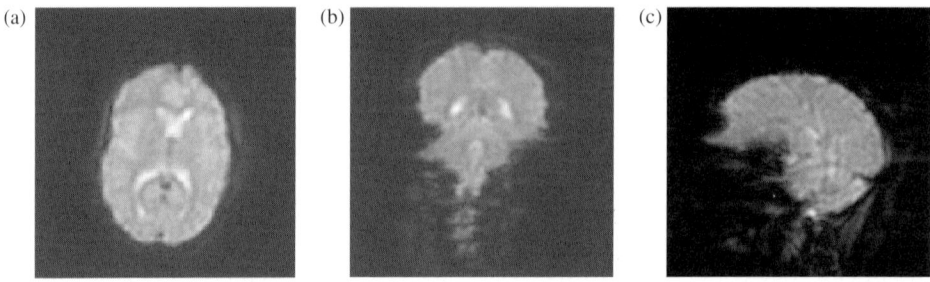

Fig. 2.2. Geometric distortion from Cartesian EPI data: (a) axial image; (b) coronal image; (c) sagittal image. Distortion is least for the axial image because of the solenoidal design of the axial gradient coils which cannot be used for the other two directions.

phase encode direction because of the relatively long T. Equation (2.1) may be expressed in terms of a position dependent PSF, or kernel, H as

$$\rho_1(\vec{r}') = \int_{\mathcal{A}} \rho(\vec{r}) H(\vec{r}', \vec{r}) \, d\vec{r}, \tag{2.2}$$

where $\vec{r} = (x, y)$ and $\mathcal{A} \subset \mathbb{R}^2$ is the support of ρ. Summing up, for Cartesian EPI, magnetic field inhomogeneity causes [334]:

- A geometrical distortion, proportional to the frequency offset $\Delta\omega = \gamma \Delta B$ caused by the inhomogeneity ΔB. (Here γ is the gyromagnetic ratio of the proton, $\gamma = 2.68 \times 10^8 \, \mathrm{T^{-1} \, s^{-1}}$.)
- Blurring in the phase encoding direction.
- Ghosting in the phase encoding direction caused by extra peaks in the PSF in the phase encoding direction.
- Blurring of the ghosts, caused by spread in the secondary PSF peaks.

The geometric distortion is least for axial images (compared to sagittal or coronal images) because of the presence of higher magnitude, higher order magnetic field concomitant with the linear gradient field for the nonaxial[†] gradient fields (see Fig. 2.2). These higher magnitude nonlinear components of the nonaxial gradient fields are caused by the different construction of the nonaxial and axial gradient field coils necessary to allow access to the bore of the MRI. Modeling of the gradient fields provides a way of correcting the extra distortion caused by the concomitant field associated with the nonaxial gradient fields [124]. Non-Cartesian EPI methods, like spiral imaging, avoid most of the geometric distortion problems with the distortion effects being generally replaced by local blurring.

From Equation (2.1) it is clear that knowledge of the magnetic field inhomogeneity could be used to correct the geometric distortion of Cartesian EPI images

---

† The axial gradient field is usually labeled as the z gradient field.

either in the image space directly through interpolation [235] or before Fourier transformation, by correcting the phase of the raw data[†], determining the actual k-space locations of the raw data and then interpolating the data in k-space to a regular grid [369]. By simultaneously correcting phase and amplitude, a simultaneous correction of the associated intensity distortion may also be done [88]. Knowledge of magnetic field inhomogeneity can be obtained from the measurement of a field map using sequences with multiple echo times, T_E, that allows the phase evolution in each voxel to be measured[‡]. The field map may be measured using an EPI sequence [369] or with a properly timed spin warp spin echo[§] sequence [393]. As an alternative to the separate acquisition of a field map and the image data, it is possible to build a pulse sequence that acquires enough data for both the field map and the image simultaneously [84, 85]. Such a sequence[¶] provides an undistorted image without postprocessing when the necessary reconstruction procedures are applied on-line within the sequence and would provide a field map for every image instead of one field map for the whole fMRI time-series. Using one map for the whole series has the disadvantage that small head motions can significantly alter the susceptibility part of ΔB. Nevertheless, geometric distortions based on one field map lead to better registration of the activation map with the spin echo anatomical map [230] and transformation of the data to a standard anatomical space [111] (see Section 2.5).

An alternative approach to using a field map to correct geometric distortion is to measure the PSF H of Equation (2.2), using an appropriate pulse sequence, at a sufficient number of voxel locations (x, y) and invert Equation (2.2) [334]. Geometric distortion methods that rely on the field map, unlike PSF methods, can be affected by phase unwrapping problems, partial volume errors[‖], and eddy current errors [471]. PSF geometric distortion correction is also better able to correct for intensity distortion over field map based methods.

Without a field map (or PSF map) the average fMRI investigator is forced to accept the positioning error that results from overlaying a geometrically distorted EPI-based activation map on an undistorted high resolution spin warp anatomical

---

[†] Raw MRI data, before image reconstruction, are complex with an amplitude and phase.

[‡] The echo time, T_E, in a pulse sequence is the time from the initial 90° RF pulse to the middle of the data acquisition period.

[§] With spin warp spin echo sequences, one line of k-space is acquired per acquisition as opposed to all lines in an EPI sequence. Spin echo spin warp T_1 weighted images are commonly used as high resolution anatomical underlays for the activation maps.

[¶] An EPI sequence that measures field and image simultaneously would be of enormous use to fMRI investigators. Unfortunately such sequences are not yet supplied as standard equipment on commercial MRIs.

[‖] The term "partial volume effects" refers to the effect that the signal from one slice (really a slab) has on the signal from another slice due to the overlapping thickness profiles.

map. However, a couple of methods have been devised to correct the geometric distortion without field maps. One method requires an additional spin echo EPI image that has the same contrast as the spin warp image [412]. Then a nonlinear transformation is computed to allow the spin echo EPI image to be warped onto the spin warp image. The resulting nonlinear transformation may then be used to correct the geometric distortion in the gradient echo EPI time-series images. Another method uses models of the susceptibility-by-movement interaction to derive a field map on the basis of how the EPI distortion changes with head motion [13]. The motion information for that last method is obtained from the rigid body motion parameters obtained by aligning each EPI image in the time-series to a reference image in the time-series (see Section 2.3).

After the images have been corrected for geometric distortion, there will still be a nonuniform distribution of signal strength, typically with reduced signal near the regions of large susceptibility gradients near the sinuses. Geometric distortion can also cause an increase in signal caused by a many-to-one voxel mapping in regions of high distortion. Such regions can be identified by comparing homogeneity corrected EPI images to uncorrected images using a t-test [278] and using the resulting difference maps in the interpretation of the fMRI results.

2.3 Image alignment and head motion

A typical fMRI experiment requires a time-series of $\sim$100 volume sets of images taken $\sim$2 s apart. The person being imaged is instructed to keep his head still (or the head is restrained by some means) but some small motion is inevitable at the sub voxel level ($\sim$2 mm in the slice plane). The analysis of the time-series assumes that each voxel is fixed and that intensity change within a voxel is caused by changes at that fixed spatial location. Small movements can shift a fraction of a given tissue type in and out of a given voxel leading to intensity changes in the voxel caused by motion that could lead to the computation of a false positive activation [202], see Fig. 2.3. An obvious way around this small motion problem is to align all of the images in the time-series to one reference image in the series[†].

Aside from pulsing activity occurring within the brain due to the cycle of blood pressure caused by the beating heart, the head and brain may be considered as a rigid object. So to align the images requires that only the six rigid body motion parameters, three rotations: yaw, pitch and roll plus three translations, are needed to specify the alignment. Specifically, if $\vec{r} = [x\ y\ z]^T$ are the image-based spatial coordinates of a given brain structure in the image to be aligned and $\vec{r}' = [x'\ y'\ z']^T$

[†] Usually the first image of the time-series is not chosen because a few T_R repeats are necessary to bring the pixel intensity to a constant level that depends on the the values of the spin lattice relaxation rate, T_1, and T_R.

threshold T [95], where

$$[I] = \frac{1}{N} \begin{bmatrix} \sum(x_i - \alpha)^2 & \sum(x_i - \alpha)(y_i - \beta) & \sum(x_i - \alpha)(z_i - \gamma) \\ \sum(y_i - \beta)(x_i - \alpha) & \sum(y_i - \beta)^2 & \sum(y_i - \beta)(z_i - \gamma) \\ \sum(z_i - \gamma)(x_i - \alpha) & \sum(z_i - \gamma)(y_i - \beta) & \sum(z_i - \gamma)^2 \end{bmatrix}$$

(2.5)

with N being the number of voxels having an intensity greater than T and

$$\begin{bmatrix} \alpha \\ \beta \\ \gamma \end{bmatrix} = \frac{1}{N} \begin{bmatrix} \sum x_i \\ \sum y_i \\ \sum z_i \end{bmatrix}$$

(2.6)

being the center of gravity vector. Alignment based on such fiducial marker alignment is a true one-step computation and therefore faster by factors of 2–5 than the nonlabel-based approaches; however the accuracy of registration deteriorates with increased image noise [96]. Another approach, which might be considered as an intermediate between label-based and nonlabel-based approaches, is to align edge detected images on the basis of maximizing the correlation between the edge detected images [47]. Such contour-based registration methods have an advantage over the MSE minimization methods in that contour registration is not affected by intensity variation caused by nonmotion sources.

The nature of the MRI signal means that a simple rigid body transformation cannot remove all the effects of motion on the image. There are two main effects that a rigid body transformation cannot correct. One is the partial volume effect between slices caused by the intensity profile in the slice thickness direction. A small motion in the slice thickness direction can lead to a large change in the signal from the edge of the slice due to the large profile gradient there. The second effect may be called unintentional spin-tagging, or a spin history effect, and is the result of the first effect in combination with the history of movement. A proton spin near the edge of a slice will, in effect, experience a variable T_R that will lead to varying amounts of longitudinal magnetization recovery between images which, in turn, lead to a motion-related signal intensity variability from a fixed physical location that remains after image alignment. One solution is to model the proton magnetization history on the basis of the computed time-series of rigid body motion parameters obtained by rigid body alignment [160]. The resulting magnetization history model can then be used to adjust the pixel intensity of the individual time-series frames. Although perfect adjustment is not possible, because the original alignment depends on intensity which is adjusted after magnetization history modeling, a considerable improvement in sensitivity to BOLD signal can be had from the alignment–intensity adjustment process [160].

With longer T_R (~4 s) the effects of magnetization history become negligible because the proton spins have had the opportunity to almost completely relax in the longitudinal direction. However, motion history can still be represented in the intensity time course of a given voxel in the realigned data because of interpolation errors from the method used to actually realign the images. The most accurate interpolation method would involve, for each voxel, an interpolation, such as sinc function interpolation, that depended on the value of all the other voxels. Such approaches to interpolation can be computationally intensive [203] and can be drastically reduced by interpolating in the Fourier domain where translation in the image domain is converted to a simple phase multiplication in the Fourier domain [128]. Interpolation in the Fourier domain is particularly attractive for MRI data because the original data are collected in the Fourier domain. No interpolation method is perfect and there will be residual intensity variation caused by motion. This remaining residual variation can be covaried out by removing signal that is correlated with the motion as determined by the computed time course of the rigid motion parameters [198]. This last adjustment may be made as a preprocessing step, or it may be implemented as confound variables in a general linear model analysis of the data (see Section 4.1). Linear combinations of low frequency sinusoids are frequently used to model the remaining motion variation.

Even with extensive modeling of the head motion and its effects, there remains susceptibility caused signal variation due to the motion of objects, like the heart and lungs, that are outside the field of view of the head [466]. These variations may be addressed by methods described in Section 2.4.

2.4 Physiological artifact and noise removal

There are many sources of physiological "noise" in the fMRI signal. These sources primarily include heartbeat and breathing, which in turn affect the flow of cerebral spinal fluid (CSF) and localized pulsing in the brain. Speaking and swallowing can also be sources of systematic signal change. If the effect of these systematic sources is not accounted for, the statistical power of the subsequent analysis of the BOLD signal will be reduced. The systematic non-BOLD signal can be modeled as "nuisance variables" in a GLM or, as described in this section, the systematic signals may be removed as a preprocessing step.

2.4.1 Drift

Before physiologic effects can be measured and removed it is necessary to remove nonspecific slow signal drift from the intensity time-series. The source of non-specific signal drift includes CSF flow and spontaneous low frequency fluctuations that have been hypothesized to be related to nonspecific neural activity [253]. Small drifts in the MRI's main field and/or receiver frequency can also cause drift; low frequency drift has been observed at the edges of the brain and in cortex infoldings in cadavers [402]. Drift may be quantified by linear and higher order polynomial fits to the time course for each 3D voxel [26] or by cubic (or other) splines. With splines, a choice needs to made on the number of basis functions with more basis functions being able to model higher frequency variance. Wavelets (see Section 2.4.5) may also be used to remove a drift, or more generally, a global effect by assuming that the $\log_2(n) - 1$ scale level (for an n length time-series) represents the global signal, replacing that level with a constant level and then using the inverse wavelet transform to produce a detrended time-series. In a comparison of linear, quadratic, cubic, cubic spline and wavelet-based detrending it was found that cubic spline detrending resulted in time-series with the highest significance (p values) after a GLM analysis (see Section 4.1) [417]. The cubic spline was followed by quadratic, linear, cubic and wavelet, in order of significance with the order of cubic and wavelet switching, depending on the subject imaged. That same study also examined an auto-detrending method that chose the best (based on final p value) of the five listed detrending methods on a voxel by voxel basis in each subject.

2.4.2 Cardiac and respiratory signal

After the effects of rigid body head motion and low frequency drift have been accounted for, the next systematic effects to remove are those due to physiological processes like breathing and heart beat. The changing volume of air in the lungs causes varying magnetic susceptibility gradients across the brain, and internal motion in the brain in the form of pulsation in arteries, veins and cerebrospinal fluid is caused by the heart. The amplitude of these physiological variations increases with higher main field strength because of the increase in the corresponding susceptibility induced magnetic field gradients.

The removal of variation due to breathing and heart beat requires knowledge of the breathing and heart beat time courses. Breathing can be measured directly using a pneumatic bellows strapped around the subject's chest and heart beat can be measured using EEG or photoplethysmograph[†] equipment. The respiratory and cardiac time courses may be used either to correct the data after acquisition or in real time to acquire the data in a manner that reduces the variation due to breathing and heart beat. The two physiologic signals can produce ghosting effects in the acquisition phase direction, especially with 3D methods that encode the third direction in k-space instead of with spatially selective slices. The physiologic ghosting artifact can be effectively removed by gating the phase encodes using either the respiratory or cardiac signals in real time. Gating based on the cardiac signal is as effective in

[†] The photoplethysmograph relies on an LED clamped on the end of the finger.

reducing physiologic signal variation as the k-space methods discussed below, but respiratory gating does not efficiently remove physiologic variation [409].

Gating of the MRI pulse sequence at the volume acquisition point, ensuring that each volume is acquired at the same phase of the cardiac cycle, can also reduce physiological signal variation. However, the signal must then be retrospectively corrected based on an estimated T_1 value and a model of the proton magnetization history because of the resulting variable T_R [199]. Any subsequent map computation must then also take into account the variable T_R in the statistical model of the BOLD function.

Recorded cardiac and respiratory time courses may be used as confound variables in the map computation stage if a GLM (see Section 4.1) is used [105] or they may be subtracted from each 3D voxel's time course as a preprocessing step. The recorded physiologic signals are not themselves subtracted from the voxel time-series, rather the recorded signal is used to determine the phases of a model of the cardiac and respiratory effects at each imaging time point t. Specifically, the physiological noise component, $y_\delta(t) = y_c(t) + y_r(t)$, may be expressed as [185]

$$
\begin{aligned}
y_c(t) &= \sum_{m=1}^{2} a_m^c \cos(m\varphi_c(t)) + b_m^c \sin(m\varphi_c(t)), \\
y_r(t) &= \sum_{m=1}^{2} a_m^r \cos(m\varphi_r(t)) + b_m^r \sin(m\varphi_r(t)),
\end{aligned}
\tag{2.7}
$$

where the superscripts c and r refer to cardiac and respiration respectively and $\varphi_c(t)$ and $\varphi_r(t)$ are the corresponding physiologic phases. The phases are variable from cycle to cycle and the cardiac phases are easy to compute based on the time between the R-wave peaks. The respiratory phases may be computed using a histogram equalization method given by Glover *et al.* [185]. Once the phases for each imaging time point are determined, the Fourier coefficients for Equation (2.7) may be computed from

$$
\begin{aligned}
a_m^x &= \sum_{n=1}^{N} [y_x(t_n) - \overline{y_x}] \cos(m\varphi_x(t_n)) \bigg/ \sum_{n=1}^{N} \cos^2(m\varphi_x(t_n)), \\
b_m^x &= \sum_{n=1}^{N} [y_x(t_n) - \overline{y_x}] \sin(m\varphi_x(t_n)) \bigg/ \sum_{n=1}^{N} \sin^2(m\varphi_x(t_n)),
\end{aligned}
\tag{2.8}
$$

where x is c or r, $\overline{y_x}$ is the average of the time-series and N is the total number of time points [185]. Traditionally the signals of Equation (2.7) are simply subtracted from the voxel time courses [93, 185] although projecting out the signals as in Equation (2.9) may be a better idea.

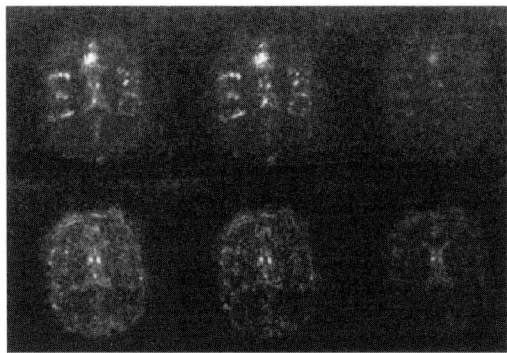

Fig. 2.4. Distribution of brain activity that is correlated with the cardiac signal (top row) and the respiratory signal (bottom row). The first column shows activity seen in the original time-series data at the cardiac and respiratory frequencies before preprocessing. The second column shows cardiac and respiratory frequency activity left in the time-series after preprocessing that subtracts the components given by Equation (2.7) as determined from using the center k-space points from a spiral sequence. The third column shows cardiac and respiratory frequency activity left in the time-series after preprocessing that subtracts the components given by Equation (2.7) as determined from pneumatic bellows and photoplethysmograph measurements. This image was taken from [185] and is used with permission.

MRI data may also be used to estimate the physiologic time courses in the absence of a separate recording. A navigator echo (essentially a stand alone unphase-encoded k-space line) may be used before each slice acquisition and the phase of the echo (center point) used to produce a time-series from which the breathing and heart beat time courses may be computed [228]. The use of a navigator has the drawback that it needs a minimum T_R to allow time for it in the pulse sequence and it is acquired at a different time than the image data that need to be corrected. Fortunately one can use the center k-space point (or small neighborhood of the center) of the actual image data, before it is reconstructed via the Fourier transform, to get the same phase time-series that the navigator echo can provide [144, 271, 460] – plus the image-based physiologic information series temporally matches the image series. Figure 2.4 compares the efficacy of removing the cardiac and respiratory components from the measured time-series using measured cardiac and respiratory rates versus using cardiac and respiratory rates as determined from central k-space data.

The determination of the breathing and heart beat time courses from center k-space data requires that multislice data be arranged in temporal, rather than slice order. If the images are not temporally ordered, the physiologic signals will be undersampled and impossible to determine because of the resulting aliasing effects [144]. The typical respiratory frequency range is from 0.1 to

0.5 Hz (6–30 cycles/min) and the cardiac frequency range is from 0.75 to 1.5 Hz (40–90 beats/min) [93]. Inherent in using the center k-space phase value is the assumption that the physiologic effects are global. This assumption is more or less true for breathing but less true for heart rate variation since pulsatile motion in the brain is more local. In a study with resting volunteers it has been shown that strong cardiac effects may be seen near the medial areas of the brain, along the middle cerebral artery near the anterior lobes and the insula around the anterior cerebral artery in the medial frontal lobes, and in the sigmoid transverse and superior sagittal sinus regions [112]. Nevertheless, good results can be had with the global effect assumption. Once a global time course for the physiological (and other) variation is obtained, it is necessary to extract the respiratory and cardiac contributions. These contributions are obtained by first Fourier transforming the global time course $\vec{g}$ to reveal the respiratory and cardiac frequencies. The respiratory time course $\vec{r}$ is reconstructed by computing the inverse Fourier transform of $F\vec{g}$ multiplied by a band pass centered around the respiratory frequency. The cardiac time course $\vec{c}$ is similarly computed with the selection of the proper bandpass width being somewhat empirical but necessary because respiration and heart beat are only approximately periodic. Finally a corrected time course, $\vec{D}$, for each voxel is computed from the original (post alignment) time course $\vec{d}$ by projecting out the respiratory and cardiac time courses [144]:

$$\vec{D} = \vec{d} - \frac{\langle \vec{d}, (\vec{r} + \vec{c}) \rangle (\vec{r} + \vec{c})}{\|(\vec{r} + \vec{c})\|^2}. \tag{2.9}$$

Instead of using information from the central k-space point, it is possible to produce a physiologic time-series using either the mean of the magnitude images or the mean of a low frequency region of influence (ROI) in k-space (to avoid problems with low signal to noise ratio at high spatial frequencies) [229].

2.4.3 Global signal removal

Besides MRI drift, cardiac and respiratory influences, other non-BOLD factors can systematically contribute to variation in the fMRI time-series. Speaking or swallowing in the MRI can produce main magnetic field changes of up to 0.087 ppm in the inferior region of the brain and some speech can cause additional changes of up to 0.056 ppm in the frontal region [42]. The acoustic noise from the gradient switching can also directly affect the BOLD response in auditory, motor and visual cortices [91]. Given the variable sources, some investigators choose to simply quantify the global effects instead of trying to attribute, and account for, the global effects as being due to cardiac and respiratory processes. If the global signal is correlated with the experimental paradigm, then different activation maps will be obtained if the global signal is or is not accounted for in the analysis [3]. These correlations could be introduced by task-related breathing, pain-related studies or overt speaking among other causes.

Within the framework of the GLM (see Section 4.1) two main approaches have been proposed to incorporate the measured global signal $G(t)$ (which is the time course of the image means). One is an additive term model (an ANCOVA or ANalysis of COVAriance model) that can be expressed as[†]

$$Y_i(t) = \mu_i + \alpha_i h(t) + \beta_i[c(t) + G(t)] + \epsilon_i(t), \qquad (2.10)$$

where $Y_i(t)$ is the intensity of voxel i at time t, μ_i is the average of voxel i's time course, h is the response vector of interest, c is a response (covariate) of no interest and ϵ_i represents noise [178]. The other model is a proportional multiplicative or ratio scaling model

$$Y_i(t) = G(t)[\mu_i + \alpha_i h(t) + \beta_i c(t) + \epsilon_i(y)]. \qquad (2.11)$$

The response vector h can be as simple as a binary valued function that represents the presentation time course or, more usually, a model of the hemodynamic response (see Section 4.1). The popular software package SPM offers the method of Equation (2.11) in addition to grand mean session scaling for performing between-subjects GLM analysis.

Gavrilescu *et al.* [178] compare grand mean session scaling, ANCOVA scaling, and proportional scaling with two methods of their own, a masking method and an orthogonalization method and find decreased sensitivity to activations in the three former methods compared to their own. The masking method begins by assuming that the global signal $G(t)$ can be decomposed as

$$G(t) = G_m + G_v(t) + A(t) + B(t), \qquad (2.12)$$

where G_m represents the mean of the global time course, $G_v(t)$ represents variation around that mean that is not correlated with $h(t)$, $A(t)$ is variation around the mean that is correlated with $h(t)$ because of the task induced activation and $B(t)$ represents the variation around the mean that is correlated with $h(t)$ due to other physical and physiological processes. Processes $G_v(t)$, $A(t)$ and $B(t)$ are assumed to be different in active and nonactive areas so that there should be two choices of $G(t)$, namely $G_n(t)$ and $G_a(t)$ for nonactive and active regions, for use in Equations (2.10) and (2.11). The determination of the mask that divides the image into active and nonactive regions proceeds by an iterative process. The assumption of different $G_n(t)$ and $G_a(t)$ is made to account for the effect that the activation response has on the global signal. The orthogonalization method is to use as $G(t)$ in Equations (2.10)

[†] Here the functions abstractly represent the data or design matrix vectors, for example $Y_i(t_n)$ would represent the intensity at voxel i and time point n.

and (2.11) the part of the global signal that is orthogonal to any nonconstant column of the design matrix (see Section 4.1). In other words, if $\mathcal{G}$ is the measured global signal and d_k are the nonconstant columns of the design matrix for $1 \le k \le K$, then one should use, if the d_k are mutually orthogonal,

$$G = \mathcal{G} - \sum_{k=1}^{K} \langle \mathcal{G}, d_k \rangle d_k / \|d_k\|^2 \qquad (2.13)$$

in Equations (2.10) and (2.11). In the case of nonorthogonal d_k a Gram–Schmidt orthogonalization procedure may be used in place of Equation (2.13).

Desjardins *et al.* [121] give a derivation that an adjusted global signal should be used in the proportional scaling model, Equation (2.11), where the adjusted global signal is defined by the measured global signal multiplied by its time course mean. They show that such adjustment can account for the contribution of local BOLD signals to the global signal.

Another approach is to model the global effect using the global signal, similarly to how the cardiac and respiratory signals may be modeled from the global signal (Section 2.4.2), but without explicitly identifying the source. The linear model of the global signal (LMGS) used by Macey *et al.* [293] first computes the global signal and subtracts the mean to produce an offset global signal $\vec{G}$. Then a model

$$\vec{Y}_i = a_i \vec{G} + \vec{Y}_i^{\,\mathrm{adj}} \qquad (2.14)$$

was computed for every voxel i to give an adjusted voxel time-series $\vec{Y}_i^{\,\mathrm{adj}}$ for use in subsequent activation map computation. LMGS was shown to have an advantage over spline modeling of the global effects in that high frequency global effects could be removed by LMGS but not by spline modeling.

2.4.4 Filtering techniques for noise removal

Noise, whatever its source, reduces the significance of detected activations. Parrish *et al.* [354] suggest that a map that shows the spatial variation of the signal to noise ratio[†] (SNR) is useful for showing regions where activations of a given amplitude can be detected by a given statistical procedure.

Most pulse sequences apply a spatial filter, by multiplication in k-space, to the acquired data that spatially smooths the data and increases its SNR. The type of filter used in the pulse sequence is generally selectable by the MRI operator from a set of given choices. Lowe and Sorenson [287] show, using receiver operating characteristic (ROC[‡]) methods, that the Hamming filter is the best choice, followed by a Fermi

---

[†] Signal to noise ratio is generally defined as μ/σ, where μ is the average signal and σ is the standard deviation.

[‡] ROC analyses plot specificity (false positives) versus sensitivity (true positives) while varying a parameter.

filter approximation if a Hamming filter is not available. They also note that the convolution of a Gaussian kernel with computed SPM (see Section 4.1), a commonly used smoothing approach, provides similar SNR improvements as the spatial filtering of k-space data but note that all convolution-based spatial filtering techniques can attenuate very small signals dramatically. Gaussian spatial smoothing can also be used before map computation with similar results.

Spatial filtering can also reduce the interpolation error effects from rigid body registration [292] (see Section 2.3). In the case of nonlinear alignment between subjects to a standardized brain atlas (Section 2.5), spatial smoothing becomes more important to reduce such interpolation errors and the effects of variable brain topography.

Spatial filtering can inappropriately average voxel time courses between active and nonactive voxels. A solution to this problem is to use adaptive filters. Using appropriate classification schemes, it is possible to divide voxels into classes and then define binary masks to produce subimages that can be independently spatially filtered and then combined to produce the filtered image [405]. Software for such an approach, which then avoids averaging nonactive time courses with active ones, is available[†].

Weaver [437] used a unique monotonic noise suppression technique with good results. With this technique, extrema in the image are identified along the horizontal, vertical and two diagonal directions and the data between the extrema are forced, using a least squares algorithm, to vary monotonically between the extrema.

A more sophisticated approach than adaptive filtering is to use a Markov random field (MRF) approach in a Bayesian framework. Following Descombes *et al.* [119] we outline how this approach works. The goal of the MRF approach is to preserve edges in the data both spatially, where they occur between active and nonactive regions and between gray matter and white matter, and temporally, where they occur between task and rest periods. To fix ideas, consider a 3D MRF scheme for a time-series of 2D images (the 4D approach is a straight forward extension). The set of *sites* for the MRF is then

$$S \times T = \{s = (i,j),\, t | 1 \leq i \leq I,\, 1 \leq j \leq J,\, 1 \leq t \leq T\}, \qquad (2.15)$$

where s represents the spatial coordinate and t the temporal coordinate. The *state space*, Λ, is the set of all possible intensity values (typically integers between 0 and 4095 for MRI data). An fMRI signal (data set), Y, is then a Λ-valued function on the set of sites. The set of all possible data sets Y is known as the *configuration space* Ω. An MRF is a probability distribution on Ω that satisfies a reflexive condition on finite neighborhoods in the site space and, through the Hammersley–Clifford theorem, it can be shown that an MRF can be written as a Gibbs field where the

---

[†] The software may be found at http://www.ece.umn.edu/users/guille/.

probability distribution, P, is defined as

$$P(Y) = \frac{1}{Z} \exp(-U(Y)) \tag{2.16}$$

for all $Y \in \Omega$, where Z is a normalization factor (to ensure that the total probability is 1) known as the partition function, and U is an energy function that consists of a sum over *cliques*, c, (a clique is a finite subset of sites) of potential functions. The MRF-based distributions are used within a Bayesian framework to determine the adjusted ("smoothed") data Y from the given raw data X in Ω under the assumption that $X = Y + \eta$, where η represents noise. The Bayesian approach is based on Bayes law, which states

$$P(Y|X)P(X) = P(X|Y)P(Y), \tag{2.17}$$

the probability of Y given X times the probability of X equals the probability of X given Y times the probability of Y. From Equation (2.17) we have that the a posteriori probability, $P(Y|X)$, is given by

$$P(Y|X) \propto P(X|Y)P(Y), \tag{2.18}$$

where $P(X|Y)$ refers to the likelihood model (or data driven term or goodness of fit) and $P(Y)$ refers to the prior model. The MRF model is used to define the prior model and a likelihood model is typically based on Gaussian probability distributions. For application to fMRI, Descombes *et al.* [119] use a Φ-model (see below) for the likelihood model in place of a Gaussian distribution to handle noise outliers caused by physical and physiologic processes. Putting the MRF together with the noise likelihood model gives (as shown in [119])

$$P(Y|X) \propto \exp(-U(Y)), \tag{2.19}$$

where, using $Y = \{y_{(i,j)}(t)\}$ and $X = \{x_{(i,j)}(t)\}$,

$$U(Y) = \sum_{(i,j),t} V_L(y_{(i,j)}(t))$$

$$+ \sum_{c=\{t,t+1\}} \sum_{(i,j)} V_T(y_{(i,j)}(t), y_{(i,j)}(t+1))$$

$$+ \sum_{c=\{(i,j),(i+1,j)\}} \sum_{t} V_I(y_{(i,j)}(t), y_{(i+1,j)}(t))$$

$$+ \sum_{c=\{(i,j),(i,j+1)\}} \sum_{t} V_J(y_{(i,j)}(t), y_{(i,j+1)}(t)), \tag{2.20}$$

where the potentials are given by

$$V_L(y_{(i,j)}(t)) = \frac{-\beta_L}{1 + (y_{(i,j)}(t) - x_{(i,j)}(t))^2/\delta^2} \tag{2.21}$$

for the likelihood model and

$$V_T(y_{(i,j)}(t), y_{(i,j)}(t+1)) = \frac{-2\beta}{1 + (y_{(i,j)}(t) - y_{(i,j)}(t+1))^2/\delta^2}, \tag{2.22}$$

$$V_I(y_{(i,j)}(t), y_{(i+1,j)}(t)) = \frac{-A\beta}{1 + (y_{(i,j)}(t) - y_{(i+1,j)}(t))^2/\delta^2}, \tag{2.23}$$

$$V_J(y_{(i,j)}(t), y_{(i,j+1)}(t)) = \frac{-\beta}{1 + (y_{(i,j)}(t) - y_{(i,j+1)}(t))^2/\delta^2} \tag{2.24}$$

for the prior model. The parameter A is equal to the aspect ratio of the acquisition k-space matrix (equal to 1 for square acquisition matrices) to account for differences in resolution in the i and j directions, $\beta_L = 1$, because of the unspecified proportionality constant in Equation (2.19), and β and δ are model parameters to be chosen[†]. The potentials of Equations (2.21)–(2.24) are known as Φ-models because they are based on the function

$$\Phi(u) = \frac{-\beta}{1 + (|u|/\delta)^p} \tag{2.25}$$

and $p = 2$ is used so that the Φ-model approximates Gaussian behavior near the origin. The desired adjusted data Y are obtained by maximizing the a posteriori probability (MAP – see Equation (2.19)) which Descombes et al. [119] do by using simulated annealing. Based on experiments with fMRI data, Descombes et al. [119] suggest that $\delta = 4$ and $\beta = 0.4$ provide good results.

Kruggel et al. [257] compare a number of different filtering schemes in the temporal dimension for their ability to correct for baseline fluctuation or to restore the signal. For baseline (low frequency) correction the following methods were investigated.

1. Moving average (MA) filter. This filter computes the mean in a window of length $2N + 1$:

$$y_s^f(t_0) = \sum_{r=-N}^{N} y_s(t_0 + r)/(2N + 1), \tag{2.26}$$

where here, as below, y^f represents the filtered time-series and y represents the original data. A window length of $\sim$1.4 times the length of a single trial was found to be optimal.

2. Finite impulse response low pass (FIR-LP) filter, where

$$y_s^f(t_0) = \sum_{r=-N}^{N} \phi_r w_r \, y_s(t_0 + r), \tag{2.27}$$

[†] Criticism of Bayesian approaches usually stems from the lack of fixed rules available to select parameters like β and δ in a prior model.

where ϕ_r denotes $2N + 1$ low pass filter coefficients for a cutoff frequency λ (taken to be less than the stimulation frequency ν) defined by

$$\phi_r = \begin{cases} \lambda/\pi & \text{if } r = 0, \\ \sin(r\lambda)/\pi & \text{if } 0 < |r| \leq N \end{cases} \tag{2.28}$$

and w_r are the $2N + 1$ Hamming window coefficients

$$w_r = 0.54 + 0.46\cos(\pi r/N), \quad -N \leq r \leq N. \tag{2.29}$$

In other words the filter is a product of sinc and Hamming filters.

3. Autoregressive (AR) filter:

$$y_s^f(t) = \sum_{r=1}^{p} \alpha_r\, y_s(t - r) + \epsilon(t), \tag{2.30}$$

where α_r is the p parameter of the AR(p) process [440] and $\epsilon(t) \sim N(0, \sigma^2)$. It was found that $p \geq 15$ was needed to produce good results.

4. Stateless Kalman filter:

$$y_s^f(t) = y_s(t - 1) + (p/(p + r))(y_s(t) - y_s(t - 1)), \tag{2.31}$$

where $p = (1 - p/(p + r))p + q$ is the estimated covariance, q is the process noise and r is the measurement covariance. Kruggel *et al.* [257] set $r = 0.01, q = r/10^f$ and $f \geq 1$ to get reasonable performance.

Of these baseline filtering methods, the MA filter performed best but the Kalman and FIR-LP filters were less sensitive to changes in filter parameters. The AR filter was judged as too computationally intense and too nonlinear to be of much use. For signal restoration (high frequency filtering) Kruggel *et al.* [257] consider the following temporal filtering methods, applied after the FIR-LP filtering (at $\lambda/\nu = 1.5$).

1. FIR-LP filtering. An optimum at $\lambda/\nu = 0.38$ was found.
2. Temporal Gaussian filter:

$$y_s^f(t_0) = \sum_{r=-N}^{N} g_r\, y_s(t_0 + r), \tag{2.32}$$

where

$$g_r = \frac{\exp(-r^2/2\sigma^2)}{\sqrt{2\pi}\sigma}, \quad -N \leq r \leq N. \tag{2.33}$$

Selection of $\sigma \geq 1.6$ and $N \geq 3\sigma$ gave the best results. See [46] for another application using Gaussian windows.

3. AR filters with $p = 2$ gave good results.
4. Spatial Gaussian filtering. This popular method was not found useful for restoring small signals in event-related fMRI (see Chapter 3).
5. MRF restoration as described above [119].

Of the five methods listed here, MRF performed the best.

Functional data analysis (FDA) techniques may be used to extract a smooth underlying function from the discrete fMRI data set [368]. The smoothed data may then be subjected to the same, or similar, analysis that would be applied to raw data. Working with slices, Godtliebsen *et al.* [186] construct a smooth function representation of the data on $[0, 1] \times [0, 1] \times [0, 1]$ of a slice time-series, specifically modeling the data, z_{ijk} as

$$z_{ijk} = \rho(x_i, y_i, t_k) + \epsilon_{ijk}, \tag{2.34}$$

where ϵ_{ijk} represents independent and identically distributed (iid) random noise with zero mean and variance σ^2, $x_i = i/n$, $y_j = j/n$, $t_k = k/m$ for an $n \times n$ image taken at m time points and ρ is the underlying smooth function. Given two kernel functions, K and L, defined on $[0, 1]$, the estimate $\hat{\rho}(x_p, y_q, t_r)$ of ρ at the given time points is given by

$$\hat{\rho}(x_p, y_q, t_r) = B_{pqr}/A_{pqr}, \tag{2.35}$$

where

$$B_{pqr} = \frac{1}{mn^2} \sum_{i=1}^{n} \sum_{j=1}^{n} \sum_{k=1}^{m} K_{pi}^h K_{qj}^h K_{rk}^h L_{pqij}^g z_{ijk}, \tag{2.36}$$

$$A_{pqr} = \frac{1}{mn^2} \sum_{i=1}^{n} \sum_{j=1}^{n} \sum_{k=1}^{m} K_{pi}^h K_{qj}^h K_{rk}^h L_{pqij}^g \tag{2.37}$$

with

$$K_{pi}^h = K^h(x_p - x_i), \tag{2.38}$$

$$K_{qj}^h = K^h(y_q - y_j), \tag{2.39}$$

$$K_{rk}^h = K^h(t_r - t_k), \tag{2.40}$$

$$L_{pqij}^g = L^g(\bar{z}_{pq} - \bar{z}_{ij}), \tag{2.41}$$

where $K^h(x) = K(x/h)/h$, $L^g(x) = L(x/g)/g$ and $\bar{z}_{pq} = \sum_{k=1}^{m} z_{ijk}/m$. Godtliebsen *et al.* also give methods for estimating the bias and variance of the $\hat{\rho}(x_p, y_q, t_r)$ estimates.

LaConte *et al.* [261] use a nonparametric prediction, activation, influence, and reproducibility resampling (NPAIRS) method to evaluate the effect on performance metrics of alignment, temporal detrending and spatial smoothing. NPAIRS is an alternative to ROC analysis where results are plotted in a reproducibility–mean-prediction-accuracy plane instead of the usual false-positive–true-positive plane. It is found that there is little impact on the performance metrics with alignment, some benefit with temporal detrending and the greatest improvement was found with spatial smoothing.

2.4.5 Wavelet transform basics and wavelet denoising

The wavelet transform of a function $f \in L^2(\mathbb{R})$ begins with a two-parameter family of functions $\psi^{a,b}$, known as wavelets, which are derived from a single mother wavelet function Ψ as

$$\psi^{a,b}(x) = \frac{1}{\sqrt{|a|}} \Psi \left(\frac{x-b}{a} \right). \tag{2.42}$$

Any function Ψ for which

$$\int_{\mathbb{R}} \frac{|\mathcal{F}\Psi(\xi)|^2}{\xi} d\xi < \infty \tag{2.43}$$

qualifies as a mother wavelet. A sufficient condition that leads to the satisfaction of Equation (2.43) is that $\mathcal{F}\Psi(0) = 0$. The wavelet transform $\mathcal{W}f$ of f is given by

$$\mathcal{W}f(a,b) = \langle f, \psi_{a,b} \rangle. \tag{2.44}$$

When Equation (2.43) is satisfied, the wavelet transform is invertible; i.e. $\mathcal{W}(L^2(\mathbb{R}))$ is a reproducing kernel Hilbert space, see Daubechies's book [115] for details. Define

$$\psi_{m,n}(x) = \psi^{a_0^m, nb_0 a_0^m} = a_0^{-m/2} \Psi(a_0^{-m}x - nb_0) \tag{2.45}$$

to be a discrete set of wavelet functions indexed by the integers n and m. Then with the proper selection of a_0 and b_0 the set $\{\psi_{m,n}\}$ forms a countable basis for $L^2(\mathbb{R})$. When Ψ is formed in the context of a multiresolution analysis, defined next, we may set $a_0 = 2$ and $b_0 = 1$.

A multiresolution analysis [115] consists of a succession of subspaces $V_j \subset L^2(\mathbb{R})$ such that

$$\cdots V_2 \subset V_1 \subset V_0 \subset V_{-1} \subset V_{-2} \cdots \tag{2.46}$$

with[†]

$$\overline{\bigcup_j V_j} = L^2(\mathbb{R}) \quad \text{and} \quad \bigcap_j V_j = \{0\} \tag{2.47}$$

plus an orthonormal basis for each subspace V_j that is obtained by dilation of the basis of V_0 which, in turn, is given by translations of a mother scaling function Φ. Specifically, if we let

$$\phi_{j,n}(x) = 2^{-j/2} \Phi(2^{-j}x - n), \tag{2.48}$$

then $\{\phi_{j,n}\}$ forms an orthonormal basis for V_j for fixed j. Projection of a function f into V_j is said to be a representation of f at scale j. We make use of this idea by

---

[†] The overbar in Equation (2.47) represents closure.

associating a function $P_0 f$ with a given data vector $\vec{y}$ as follows:

$$P_0 f(x) = \sum_n y_n \phi_{0,n}(x), \qquad (2.49)$$

where P_0 is the projection into V_0 operator. In other words, we set $y_n = \langle f, \phi_{0,n} \rangle$ somewhat arbitrarily. With a multiresolution setup, it is possible to construct a wavelet basis $\{\psi_{j,n}\}$ for the subspaces W_j, where

$$V_{j-1} = V_j \oplus W_j, \qquad (2.50)$$

which allows us to write

$$P_{j-1} f = P_j f + \sum_k \langle f, \psi_{j,k} \rangle \psi_{j,k}. \qquad (2.51)$$

Equation (2.51) gives a decomposition of a high resolution $P_{j-1} f$ into a lower resolution version (low pass filtered version) $P_j f$ plus difference information (high pass filtered version) given by $\sum_k Wf(2^j, n2^j) \psi_{j,k}$. This process may be iterated J times to give a decomposition of $P_0 f$ into J wavelet levels plus a low resolution version $P_J f$ which is characterized by "approximation coefficients"

$$a_{j,k} = \langle f, \phi_{j,k} \rangle. \qquad (2.52)$$

The quantities

$$d_{j,k} = \langle f, \psi_{j,k} \rangle \qquad (2.53)$$

are known as the wavelet coefficients (or "detail coefficients") and can be computed using fast numerical methods [363] for a given data vector $\vec{y}$. The wavelet transform is invertible in the sense that given a set of wavelet coefficients plus the low pass representation of the original data at scale J, one can recover $\vec{y}$ exactly. Note that there are many ways to choose the wavelet and scaling function pairs, Ψ and Φ, with the expected form of the data determining, in many cases, which pair is likely to be the most useful. By relaxing the orthonormal basis condition, wavelet transforms may be formulated in which one pair of scaling and wavelet functions is used to decompose the data and another pair is needed for reconstruction (the inverse wavelet transform).

The approach given above for functions on $L^2(\mathbb{R})$ can be extended using tensor products to give wavelet transforms for functions on $L^2(\mathbb{R}^n)$. The case $L^2(\mathbb{R}^2)$ corresponds, of course, to images and an example of a numerical wavelet decomposition of an image is given in Fig. 2.5. Generalizations of the wavelet transform are possible with the wavelet packet transform being one of the more common. In a wavelet packet transform, each wavelet difference function is decomposed as if it were a scaled function, according to Equation (2.51). An example of a wavelet packet transform of an image is also given in Fig. 2.5.

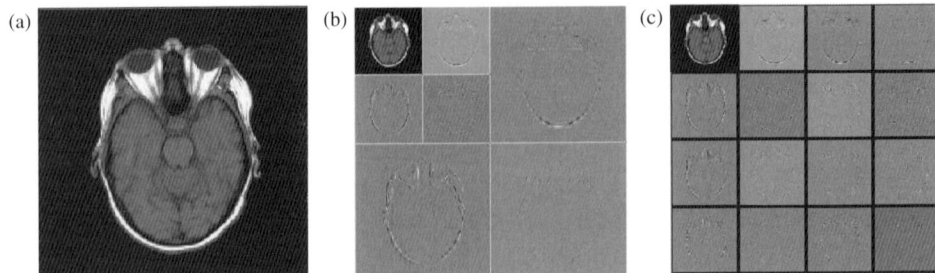

Fig. 2.5. Wavelet transforms of a 2D image: (a) original image; (b) wavelet transform to two levels. A decimating wavelet transform (using the Haar wavelet
function) is illustrated here. With a decimating wavelet transform, images at level
j have half as many coefficients (values) as the image at level $j − 1$. Decimated
wavelet transforms are numerically fast and exactly invertible. After wavelet transform to one level, four images are produced, one scale image and three wavelet
images. There is a wavelet image in the horizontal direction (bottom left image),
a wavelet image in the vertical direction (top right image) and a diagonal wavelet
image (bottom right image). The second level wavelet transform decomposes the
first level scale image into three more wavelet images and a scale image (top left-
most image). (c) Wavelet packet transform to two levels. Under the wavelet packet
transform, each of the first level wavelet images is decomposed into three wavelet
and one scale image, in addition to the first level scale image.

To accomplish wavelet denoising, the wavelet (and scale) coefficients are computed, the wavelet coefficients modified, and then the inverse wavelet transform
is applied to yield denoised data. The wavelet coefficients may be modified by
hard thresholding or by soft thresholding. In hard thresholding, any wavelet coefficient smaller than a given threshold, t, is set to zero. In soft thresholding, a given
threshold, t, is subtracted from all wavelet coefficients. Donoho [123] has shown that
soft thresholding, also known as wavelet shrinkage, represents an optimal denoising
approach given Gaussian noise characteristics. It is readily appreciated that wavelet denoising is essentially a sophisticated low pass filter technique, removing high
frequency information at multiple scale, or subband, levels. The soft thresholding
approach was the first wavelet denoising method to be applied to MRI [438].

Soft thresholding is optimal in the case that the image noise is iid as a normal
(Gaussian) distribution with mean zero and standard deviation σ ($\sim N(0, \sigma)$). For the
originally acquired complex-valued k-space data the white Gaussian noise model is
a good approximation to the noise, generated by thermal processes, found in each
of the real and imaginary channels. The images, however, are created by taking the
magnitude of the Fourier transform of the original k-space data so that the noise in
the images has a nonzero mean and a Rician (magnitude dependent) distribution
instead of Gaussian [345]. One approach, then, is to apply wavelet shrinkage to
the original k-space data [8, 470] but this approach can be compromised by phase

errors in the data [345]. Given the non-Gaussian noise distribution, the optimal wavelet filtering strategy is different from the soft thresholding one. Specifically, a Wiener type of filter works better with the filter constructed as follows. Let I be a subscript that indexes the translation (k) and dilation (j) indices, given, for example, in Equation (2.53), plus, for 2D and higher wavelet transforms, orientations so that we may write the scaling and wavelet coefficients of a signal s (image[†]) as

$$d_I = \langle s, \psi_I \rangle \quad \text{and} \quad c_I = \langle s, \phi_I \rangle \tag{2.54}$$

respectively. Then the wavelet coefficients may be filtered via

$$\hat{d}_I = \alpha_I d_I, \tag{2.55}$$

where the inverse wavelet transforms of $\{\hat{d}_I\}$ and $\{c_I\}$ give the wavelet filtered image. The Wiener type filter suggested by Nowak [345] is given by

$$\alpha_I = \left(\frac{d_I^2 - 3\sigma_I^2}{d_I^2} \right)_+, \tag{2.56}$$

where $(\cdot)_+$ means set the answer to zero if the term in brackets is negative and σ_I is an estimate of the noise that can be estimated by measuring the square of background noise surrounding the head, which will have a mean of $2\sigma^2$. With orthogonal wavelet transforms $\sigma_I = \sigma$. The application of Equation (2.56) to magnitude MRI images works well if the SNR is high where the Rician distribution becomes approximately Gaussian but it is better to apply the filter of Equation (2.56) to the wavelet transform *square* of the magnitude image, subtract $2\sigma^2$ from the scale coefficients (to correct for bias), compute the inverse wavelet transform and take the square root to end up with the denoised MRI image [345].

Wink *et al.* [444] use wavelet shrinkage denoising and compare activation maps computed from wavelet denoised images to those computed from Gaussian smoothed images. They find that the greater the amount of smoothing, the more false positives are generated. So wavelet denoising that produces less smooth images produces less false positives than Gaussian smoothed images or smoother wavelet denoised images. They also find that the noise in the difference of two images containing Rician noise is Gaussian.

Instead of applying wavelet denoising to the images in an fMRI time-series, the individual voxel time-series, as a temporal sequence, may be denoised. In the temporal direction the noise is correlated and not independent, however, Alexander *et al.* [9] argue that at each scale of a wavelet transform the noise between wavelet coefficients is nevertheless uncorrelated and the level of the noise at scale j may be

---

[†] An image may be represented as a vector by stacking the columns on top of one another.

estimated by

$$\sigma_j^2 = \left[\frac{\mathrm{MAV}_j}{0.6745} \right], \tag{2.57}$$

where MAV_j denotes the median absolute value of the wavelet coefficients at scale j. The noise estimate of Equation (2.57) may then be used in Equation (2.56) to produce a denoised temporal series for each voxel independently.

The standard wavelet transform involves a decimation at each scale (see Fig. 2.5) that makes the transform nontranslation-invariant. There are two ways to recover translation invariance, one is to shift the image or signal through all possible cyclic permutations and average the resulting wavelet transforms, the other is to use a wavelet transform that does not involve decimation, known as the stationary wavelet transform (SWT), to produce a redundant set of wavelet coefficients. Using the SWT, LaConte *et al.* [260] apply the following Wiener like filter for use in Equation (2.55) for denoising temporal fMRI voxel time-series that is organized into K epochs with N samples per epoch (a blocked design, see Chapter 3):

$$\alpha_{j,n} = \left(\frac{(K/(K-1))(\tilde{d}^j[n])^2 + (1/K - 1/(K-1)) \sum_{k=0}^{K-1} (d_k^j[n])^2}{\frac{1}{K} \sum_{k=0}^{K-1} (d_k^j[n])^2} \right), \tag{2.58}$$

where

$$\tilde{d}^j[n] = \frac{1}{K} \sum_{k=0}^{K-1} d_k^j[n]. \tag{2.59}$$

A more complex method of wavelet filtering may be derived using a leave-one-out cross-validation technique [340]. This method again applies to a blocked design with K epochs of N samples each. First compute the SWT $\{d_k^j\}$ of each epoch k independently. Then construct the leave-one-out estimates

$$\hat{d}_p^j[n] = \frac{1}{K-1} \sum_{k \neq p} d_k^j[n]. \tag{2.60}$$

Apply wavelet shrinkage to each epoch k time-series to leave

$$\tilde{d}_p^j[n] = \hat{d}_p^j[n](1 - \lambda^j[n]), \tag{2.61}$$

where, using $\gamma(\theta) = \min(\theta, 1)$,

$$\lambda^j[n] = \gamma \left(\frac{(\hat{d}_p^j[n])^2 - d_p^j[n]\hat{d}_p^j[n]}{(\hat{d}_p^j[n])^2} \right). \tag{2.62}$$

Average the coefficients obtained for each p from Equation (2.61) and then inverse SWT to obtain the denoised time-series. A leave-q-out version is also possible [340].

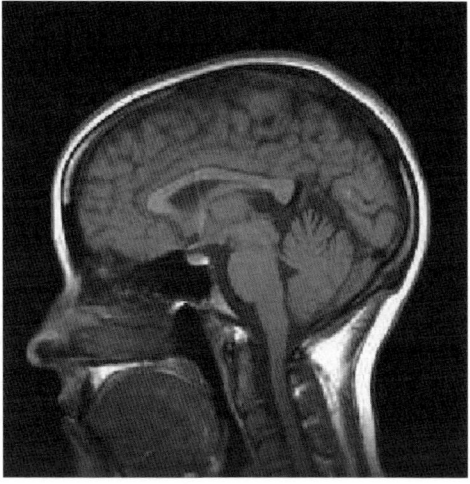

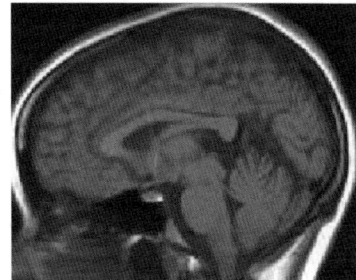

image after transformation
into the "Talairach box"

original T_1-weighted image

Fig. 2.6. An example of a transformation into Talairach coordinates. The Cartesian coordinates of the voxels in the transformed image may be used to identify specific brain structures – at least approximately. The landmarks are usually identified on a high resolution T_1-weighted "anatomical" image like the one shown here. Once the landmarks are identified, the transformation may be determined. The transformation may then be used on all images that were obtained in the same imaging session, including the EPI time-series images and computed activation maps, as long as the physical coordinates of the voxels for those images are known. The physical coordinates of the pixels can usually be determined from information saved by the pulse sequence in the image header. The geometrical distortion of the EPI images relative to the standard "spin warp" images used for identifying the anatomical landmarks introduces some inaccuracies in that approach. Those positional inaccuracies may be mollified to a certain extent by convolving the data (EPI time-series or computed activation maps) with a smoothing function like a Gaussian before transformation. The use of smoothed images in Talairach space also compensates to some degree for individual variation in brain structure when comparing activations between subjects.

2.5 Anatomical transformations to standard brain spaces

Comparing or averaging results between subjects requires that the data, in the case of transformation before analysis, or the activation maps, in the case of transformation after analysis, be registered to a common stereotaxic space represented by a template. The most common template in use is that of Talairach and Tournoux [416], see Fig. 2.6. Other templates include the Montreal Neurological Institute (MNI) probabilistic template [131, 132, 133] (see Fig. 4.7 for an example display in MNI space) and the human brain atlas (HBA) [375]. A nonlinear transformation is required to map individual data into these standard brain spaces. In addition to registering to standard templates, many of the methods described below are also useful for registering fMRI data to images obtained from other modalities like PET (positron emission tomography) or SPECT (single photon emission computed tomography) images and can also be used to warp EPI data onto high resolution spin warp anatomical MRI images. Nestares and Heeger [337] use a multiresolution method (Laplacian pyramid – a wavelet decomposition) to align low resolution functional images with high resolution anatomical images. Many of the transformations are made easier if the individual data are smoothed first and smoothing can also make up for differences in individual anatomical topography for the simpler transformations. Many of the transformations used begin with an affine transformation in one way or another.

The rigid body transformation of Equation (2.3) is a special case of the more general affine transformation that allows linear stretching and squeezing in addition to translation and rotation. By augmenting 1 to the vectors $\vec{r} = [x\ y\ z]^T$, the coordinates of points in the original image, and $\vec{r'} = [x'\ y'\ z']^T$, the coordinates of points in the transformed image, we can write the general affine transformation as

$$
\begin{bmatrix} x' \\ y' \\ z' \\ 1 \end{bmatrix} = \begin{bmatrix} m_{11} & m_{12} & m_{13} & m_{14} \\ m_{21} & m_{22} & m_{23} & m_{24} \\ m_{31} & m_{32} & m_{33} & m_{34} \\ 0 & 0 & 0 & 1 \end{bmatrix} \begin{bmatrix} x \\ y \\ z \\ 1 \end{bmatrix},
\tag{2.63}
$$

where some means to constrain the transformation so that the transformation parameters may be computed is required. In the case of registering to Talairach space, a 12-patch piecewise continuous transformation is used to squeeze the brain into the "Talairach box" [107]. Landmarks are required to define the piecewise affine transformation and these need to be manually identified. These landmarks are the anterior and posterior commisures (AC and PC), top (superior) most, bottom (inferior) most, left most, right most, front (anterior) most and back (posterior) most parts of the brain.

An affine transformation may also be used to register images to a template in a nonlabel-based manner [18]. Nonlinear methods may be used for more accurate transformation into the template space. Many of the published nonlinear methods use Bayesian methods (i.e. MAP) for their solution [19, 21]. Nonlinear nonlabel-based methods include those that use Fourier type basis functions [20] or spline warps [122] to define the transformation. Label-based methods that employ automatically extracted features have also been used [94, 116]. The various approaches for registration to a template may be evaluated using some criteria that measure registration accuracy, including evaluation methods based on wavelets [122]. Using the wavelet methodology, Dinov *et al.* [122] find that the MNI nonlinear spline warp [100] provided better registration than the AIR 12- and 30-parameter polynomial warps [447, 448] and the SPM nonlinear trigonometric warping [20].

After the activation maps have been transformed onto a common brain atlas, the identification of various sulci and gyri is useful for understanding the underlying brain function. Such anatomical region identification may be done automatically[†] [425].

2.6 Attitudes towards preprocessing

The first preprocessing step of ghost and some noise removal is usually performed at the pulse sequence level and is generally done through the selection of options in setting up the pulse sequence. For example, selection of fat saturation will eliminate the chemical shift ghost and the selection of a k-space filter will eliminate some of the noise. The incorporation of such features into pulse sequences makes it easy for any investigator to apply them.

After pulse sequence level preprocessing, it is generally considered desirable to align the time-series images using rigid body transformations (known by many as "motion correction") according to the prescriptions given in Section 2.3. However, as we have seen, some studies have shown that even this step may introduce false positives. Also, if the motion is too large, alignment will not make the resulting activation map artifacts go away. So image alignment is probably most useful for a limited range of motion amplitude; if the motion is small or nonexistent, alignment can introduce false activations from interpolation error and if the motion is too large alignment cannot compensate because the interpolation of image values becomes too crude.

The step of preprocessing to remove physiologically correlated signal is best done using measured heart and respiration rates. If those measurements are not available, the best approach may be to model such "nuisance effects" in a GLM approach to detecting the activations (see Chapter 4). If the frequencies of heartbeat and breathing are incommensurate with any periods in the experiment presentation design (see Chapter 3), then it is frequently safe to ignore the presence of the background physiologic signals.

Removing effects like heartbeat, breathing and even drift as a preprocessing step can compromise the power of the subsequent statistical analysis if not done correctly. This is because, from a variance point of view, the subtracted signals are usually not orthogonal to the signals of interest. In other words, there is some shared variance between the two signals. That is why "projecting out" the unwanted signal as per the prescription of Equation (2.9) is better than simply subtracting the confound signal. A safer approach is not to remove the effects of no interest as a preprocessing step but to explicitly model the effect as a column or columns in a

---

[†] Software is available as an SPM interface.

GLM analysis (see Section 4.1). Then the variance due to each effect is accounted for properly, and automatically.

In comparing activations between subjects, it is usually necessary to transform into a common anatomical space, as discussed in Section 2.5. However, this step may be postponed until after the maps are computed (a usual practice). So, with the exception of ghost removal and k-space filtering at the pulse sequence stage, nearly all of the preprocessing steps discussed in this chapter may be considered as optional, depending on the presentation design and on the type of subsequent time-series analysis proposed to make the activity maps.

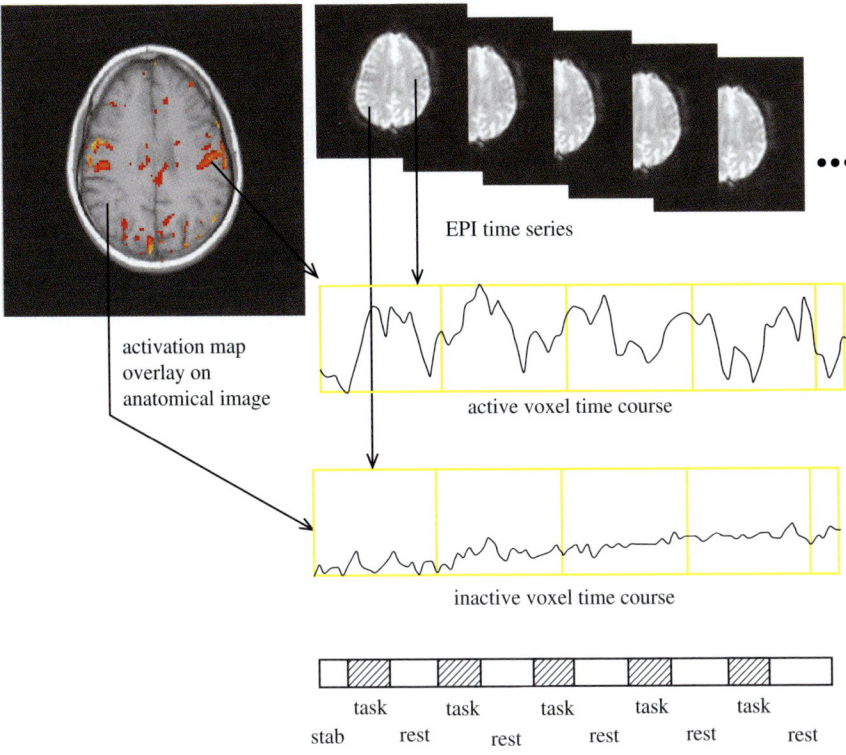

EPI time series

activation map
overlay on
anatomical image

active voxel time course

inactive voxel time course

task   task   task   task   task

stab   rest   rest   rest   rest   rest

Color Plate 1.1. Schematic of a typical EPI BOLD fMRI data set and its reduction to an activation map. An EPI time-series, represented here by a series of one slice from the volume data set at each point in time, is collected while the subject in the MRI performs a task synchronized to the image collection. The intensity of each voxel in the image set will vary in time. An active voxel will vary in step with the presented task while an inactive voxel will not. A statistical method is needed to determine if a given voxel time course is related to the task, and therefore active, or not. The active voxels are then color coded under a relevant amplitude-color relationship and the resulting activation map is overlain on a high resolution anatomical MRI image to allow the investigator to determine which brain regions are active under the given task.

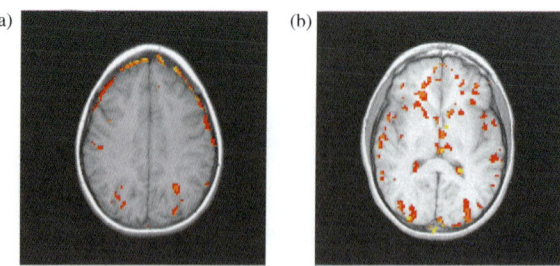
(a)   (b)

Color Plate 2.3. Small head motion in an fMRI experiment leads to false positives in activation maps. (a) The "ring artifact" is caused by motion of the edge of the brain in and out of pixels at the edge of the brain. (b) The "brain-on-fire artifact" is caused by motion perpendicular to the image plane. In both cases motion that is correlated with the task paradigm is picked up in the activation maps. Image alignment can reduce these kinds of artifacts in the activation maps if the motion amplitude is on the order of the size of a voxel or less. The identification of ring artifact is relatively straightforward. The identification of "brain-on-fire artifact", as I have called it here, can be trickier. Some tasks, such as the mental rotation task, can produce a lot of activation and false activation can be hard to identify in those cases. (The mental rotation task involves the presentation of two 3D objects in different orientations and the subject is asked to decide if the two objects are identical. To accomplish the task the objects must be mentally rotated to be compared.) Brain-on-fire artifact can be more severe than shown here but it is usually accompained by the ring artifact, especially in more superior axial slices. One hint that something is wrong is that the ventricles show activation, although this may be caused by breathing physiology. (In some investigations involving exercise, ventricle "activation" has been observed when changes in breathing were correlated with the task. It is not yet clear whether there were other motion-related false positives in those activation maps.)

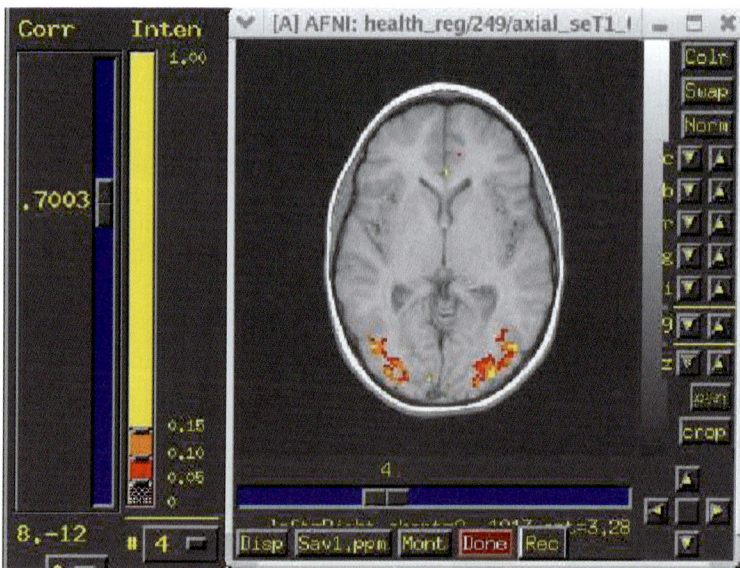

Color Plate 4.1. An example use of color-coding with an SPM. This image of the interface to the AFNI software shows a color-coded amplitude map thresholded with an SPM of correlation values. The correlation threshold in this example is set at 0.7003 by the slider on the interface and the color bar shows that the colors represent fractions of the maximum BOLD amplitude found in the data set. Many software packages for fMRI data processing have similar interfaces for displaying SPM related data.

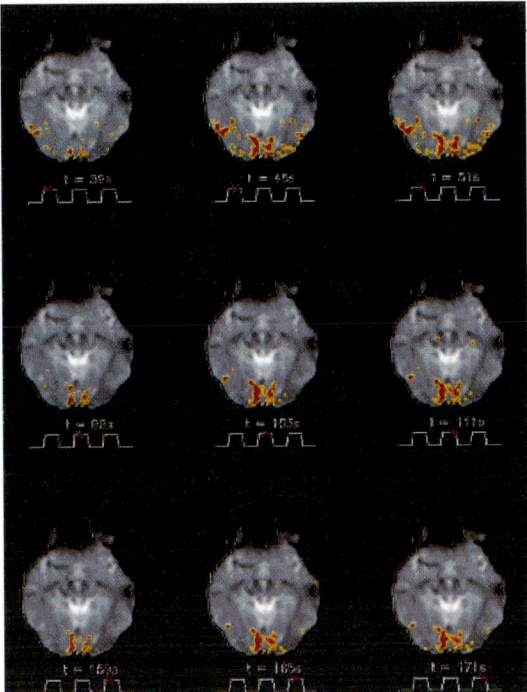

Color Plate 4.8. Example frames from an activation movie made with the random walk model of Equations (4.276) – (4.278) and a visual stimulation paradigm. Although hard to see in this reproduction, underneath the images is a representation of the presentation paradigm and a red dot on that representation showing the time of the activation map. The times, from top left to bottom right are t = 39, 45, 51, 99, 105, 111, 159, 165 and 171 s. Taken from [191], used with permission.

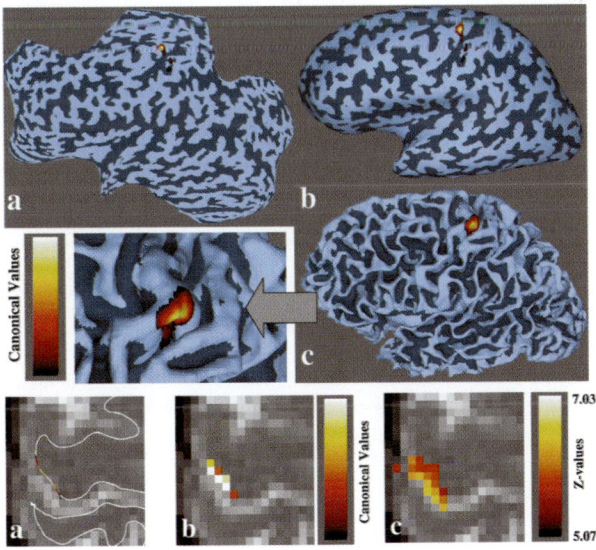

Color Plate 5.2. Activations computed on the 2D surface of the cortex using AIBFs. In part (ii): (a) shows an outline of the computed cortical surface on top of the voxels of the original EPI data along with the activation as computed on the 2D surface; (b) shows the activations as transformed by the function g of Equation (5.15) from cortical space to voxel space; (c) shows a conventional map as computed with the SPM software for comparison. Part (i) shows the same activation on: (a) the cortical flat map S_F; (b) the inflated cortical map S_I; and (c) the folded grey matter surface S_G. Taken from [247], used with permission.

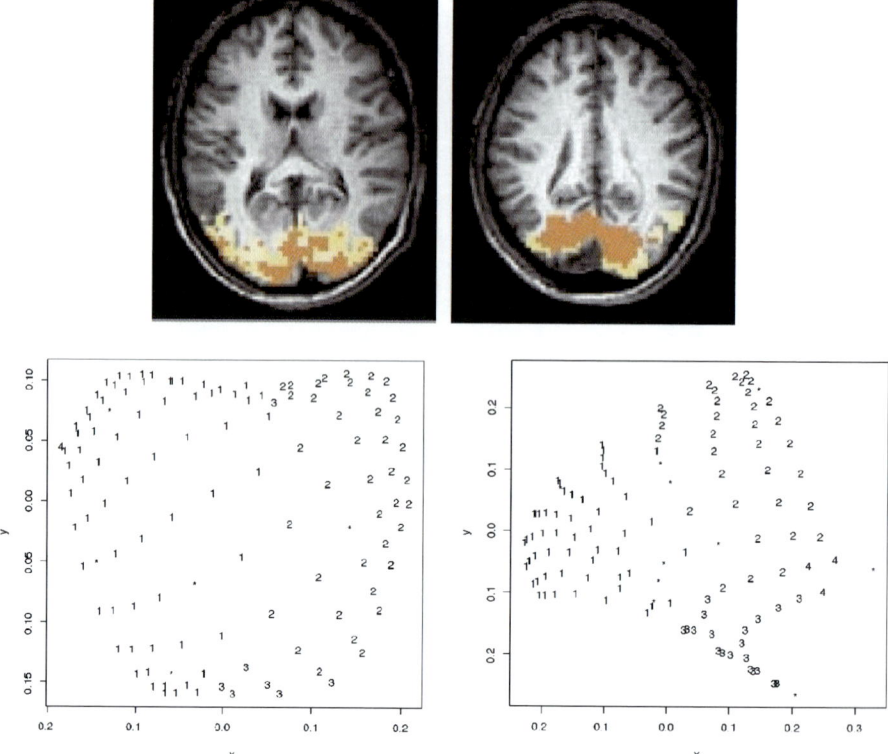

Color Plate 5.7. Example replicator dynamics solution to finding networks in a visual stimulation experiment. At the top, two slices from one subject are shown with the found networks color coded with red representing the first network and various shades of yellow representing other networks. At the bottom are the MDS maps (plotted with the two principal directions of the similarity structure matrix $[W]$ represented as the x and y directions) for each slice with the networks being represented by their numbers. In this example, SOM clustering was done first, followed by the determination of networks of cluster centers by replicator dynamics from the SOM clusters. Note how the SOM organization shows up in the MDA clusters as expected. Taken from [286], ©2002 IEEE, used with permission.

3

Experimental designs

Before an fMRI time-series can be acquired, a question needs to be posed and an experiment designed. There are two main types of experimental design in fMRI, the blocked design and the event-related design[†]. Figure 3.1 schematically shows the difference between the two designs. In the blocked design, a fixed number of multiple trials are presented in immediate succession in each block. The time between each trial is known as the stimulus onset asynchrony (SOA). Between blocks of stimulus presentation are blocks of rest. One cycle of blocked task and rest is known as an epoch. In an event-related design the tasks are presented individually, instead of in blocks, with a spacing, T, that may be variable [376]. The time interval T may be further broken down[‡] into the interstimulus interval (ISI) and stimulus duration (SD) so that $T = \text{ISI} + \text{SD}$.

Each experimental design has its advantages and disadvantages. For example, a blocked design will generally be more sensitive to detecting activations while an event-related design may be better able to characterize the BOLD response. Both types of design rely on signal averaging to remove noise when the activation maps are computed. An event-related design with constant ISI may be considered as a limiting example of a blocked design. With both types of design, the hemodynamic response function (HRF) may be considered, in a linear systems approach, as the convolution of the stimulus paradigm (a step function equal to 1 during the stimulus and 0 otherwise) and the impulse response function (IRF). The shape of the HRF tends to drive the analysis of blocked designs, while the shape of the IRF tends to drive the analysis of the event-related designs. For example, the stimulus paradigm function for the analysis of event-related data tends to be more a series of Dirac δ functions than a step function. For some designs, the distinction between blocked and event-related designs may be blurred as we shall see in the following sections.

[†] The blocked design has also been referred to as the state-related design [161].

[‡] Some investigators prefer to equate T and ISI.

(a)

(b)

(c)

Fig. 3.1. Blocked and event-related fMRI experimental designs. (a) Blocked fMRI design. The high regions represent task and the low regions represent rest. The tasks and rests are alternated periodically with the sum of one task and one rest being designated an epoch. (b) The blocked design interpreted as an event-related design. Each block of a blocked design typically contains repeated performance of the same or a similar task. To construct a regressor in the GLM design matrix to represent the model HRF, a model IRF may be convolved with the function illustrated in (a) for a conventional blocked design analysis or with the function illustrated in (b) for an event-related analysis of the blocked design. (c) A randomly presented event-related design. The convolution of this function with a model IRF produces the model HRF for use in a GLM design matrix.

3.1 Blocked designs

The simplest way to analyze a blocked design is to use a *t*-test to compare the average signal during the task to the average signal during the rest. Software for performing such a *t*-test is readily available in the Stimulate software [411]. Most fMRI analyses, however, either model or attempt to measure the HRF in some way, because of the availability of the relatively rapidly sampled time-series data.

A popular approach to blocked designs, because it is easy to program, is to have the stimulus presentations phase-locked with the image acquisition. This approach can lead to a sparsely sampled HRF and it has been demonstrated that designs that use distributed stimulus presentation can produce different activation maps from phase-locked presentation designs when both are similarly analyzed [364, 430]. Two ways of achieving distributed presentation are to vary the SOA from presentation to presentation (e.g. SOA $= 2000 \pm 150$ ms) or to use a T_R that is not an integer multiple of the SOA (e.g. SOA $= 3000$ ms, $T_R = 3150$ ms). Varying the presentation in this way leads to sampling the HRF at points not fixed by T_R and it is hypothesized that denser sampling is needed because the HRF may contain a high frequency component caused by, in part, "top-down" cognitive processing [364, 430]. The hypothesized high frequency components have been observed experimentally as being due to resting fluctuations in arterial carbon dioxide [445].

The HRF model for model-based analysis (Section 4.1) will be different depending on whether an assumed IRF, h, is convolved with a boxcar step function, χ_S, or with a series of Dirac δ functions (see Fig. 3.1(a) and (b)). The Dirac δ approach may better characterize the observed BOLD signal [311]. Alternatively, a mixed blocked/event-related design may be used where the stimuli may be arranged in a random ISI ($=$SOA) presentation within a block to be repeated at the next epoch. An analysis with $\chi_S * h$ reveals sustained responses, while an analysis with $\delta * h$ reveals responses more directly associated with the stimulus presentations [432, 439]. Figure 3.1 illustrates designs with one task type per run but multiple task types may be presented randomly through the imaging time sequence in an event-related design and later analyzed using separate models for each task type. The same principle also applies to blocked designs as long as the task type is constant within a task block and mixed task, mixed block/event-related designs have been successfully implemented [268].

What is the optimal repetition period for a blocked design, or equivalently, what is the optimal ISI for an event-related constant ISI presentation? Following Bandettini and Cox [27], we can provide a linear systems mathematical answer to this question. Let x represent the HRF and h represent the IRF so that

$$x(t) = \alpha \sum_{m=0}^{M-1} h(t - mT) + \beta + \zeta(t), \tag{3.1}$$

where α is the response magnitude, β is the signal baseline, $T = \text{ISI} + \text{SD}$, and ζ is the noise assumed to be stationary and white with variance σ^2. Note that Equation (3.1) is a $\delta * h$ type model but can be made to be essentially a $\chi_S * h$ type of model by setting $h \approx \chi_S$. By minimizing the amount of variance in the estimate $\hat{\alpha}$ of α for a fixed amount of scan time (time covered by the fMRI time-series) Bandettini and Cox find an optimal time T_{opt} for T as

$$T_{opt} = 2\frac{\mu_1^2}{\mu_2}, \tag{3.2}$$

where

$$\mu_1 = \int h(t)\, dt \quad \text{and} \quad \mu_2 = \int h(t)^2\, dt \tag{3.3}$$

which is equivalent to maximizing the expected value of $\int |x(t) - \bar{x}|^2\, dt$, where $\bar{x}$ is the average signal ($\beta + \text{constant}$). We may evaluate Equation (3.2) by assuming a standard model for the IRF, the gamma variate function

$$h_\Gamma(t; b, c) = \begin{cases} t^b e^{-t/c} & t > 0, \\ 0 & t < 0. \end{cases} \tag{3.4}$$

We then find

$$T_{\text{opt}} = \frac{2bc4^b \Gamma(b)^2}{\Gamma(2b)}, \tag{3.5}$$

where

$$\Gamma(\tau) = \int_0^\infty t^{(\tau-1)} e^{-t} \, dt. \tag{3.6}$$

Using realistic fMRI values of $b = 8.6$ and $c = 0.55$ s we find $T_{\text{opt}} = 11.6$ s. With SD $= 2$ s, so that $h = \chi_{[0,2]} * h_\Gamma(\cdot; 8.6, 0.55)$ we find $T_{\text{opt}} = 12.3$ s. With an SD of 2 s, Bandettini and Cox find an optimum T of 12 s experimentally and also show that, for any function h, an optimum T is given roughly by

$$T_{\text{opt}} = \begin{cases} 14 & \text{SD} \le 3 \text{ s}, \\ 14 + 2(\text{SD} - 3) & \text{SD} > 3 \text{ s}, \end{cases} \tag{3.7}$$

which shows that having equal times for task and rest is not optimal.

Comparison of the relative efficiency of blocked versus event-related experimental design (from a GLM point of view) depends on the contrast vector of interest (see Section 4.1.6). If the contrast vector emphasizes *detection* (e.g. single basis function, a model HRF), a blocked design is more efficient, under the assumptions of linear, task independent hemodynamic response. If the contrast vector emphasizes *estimation* (multiple basis functions, e.g. coefficients to characterize an IRF), then a (randomized) event-related design is more efficient [44, 310].

3.2 Event-related designs

Event-related fMRI was introduced by Buckner *et al.* [65], who used a nonparametric method (Kolmogorov–Smirnov statistic, see Section 4.5) for computing activation maps for both the event-related (with equal ISIs) and a comparative blocked design. Event-related designs are, however, better used with randomized ISI spacing [74] as this allows the IRFs to be sampled with varying degrees of overlap. Event-related designs also allow for the possible removal of task correlated signal change due to motion because the HRF signal rises more slowly than the motion induced signal and will result in a deduced IRF that can be distinguished from the motion signal [43].

Hopfinger *et al.* [223] studied the effect on a randomized event-related design at 2 T of the following four factors: (1) resampled voxel size after realignment, (2) spatial smoothing, (3) temporal smoothing and (4) the set of basis functions used to model the response from a sensitivity perspective. The basis functions considered included a gamma function HRF model (see Equation (3.4)) plus the temporal derivative of the HRF plus a dispersion derivative of the HRF (the derivative with respect to c in Equation (3.4)). They found optimal values at 2 mm^2 resampling voxel size, 10 mm FWHM[†] spatial smoothing and 4s FWHM temporal smoothing and a basis function set (modeled by columns in the design matrix, see Section 4.1) consisting of an HRF model and its temporal derivative[‡].

Josephs and Henson [238] provide a review of factors that affect event-related experimental designs. In particular, there can be considerable variation in the actual HRF and its linearity both between subjects and between brain regions within a single subject [4, 244]. This variation can cause a significant number of false negatives when a fixed HRF model is used in a GLM (see Section 4.1) for activation detection[§] [206]. Duann *et al.* [125] used ICA (see Section 5.2) to show that it is possible to evolve a double peaked hemodynamic response to a short stimulus burst in the V1 region of the occipital lobe. The HRF can also vary with age. Using a model of the HRF composed of piecewise sums of variable width half-Gaussians, Richter and Richter [373] were able to show that the latencies of the leading edge, peak and trailing edge all increased with age, where the latencies were defined as the midpoints of the middle three of the seven pieces of the fitted HRF model.

Using the efficiency criteria given in Section 4.1.6 for estimation it can be shown that a randomized selection of ISI is more efficient than constant ISI presentation. A truly optimal presentation with respect to Equation (4.68) can be done using genetic algorithms to search through the ISI parameter space [436] (software available as an SPM extension).

It is possible to mix more than one task type into a design, either for event-related or blocked designs and then separate the tasks at the map computation stage (see Section 4.1.5). With the introduction of multiple tasks, specialized event-related designs become possible [280, 283] that include:

- Permuted block designs. These may be constructed beginning with a blocked design containing blocked identical tasks with the tasks being different between the blocks, then modifying by permuting individual tasks between originally homogeneous blocks to make the blocks heterogeneous in terms of task type within the blocks.
- m-sequence designs [73]. With these designs, if $L = Q + 1$ is a prime number with Q being the number of tasks (1 is added for the null task), a specialized sequence of length $N = L^n - 1$ can be constructed whose estimation efficiency is better than a random design.

[†] FWHM = full width half maximum, in this case with a Gaussian kernel.
[‡] The use of a derivative models some of the nonlinearity of the HRF, see Section 4.2.
[§] Handwerker *et al.* [206] use the Voxbo software (http://www.voxbo.org) and the fmristat software (http://www.math.mcgill.ca/keith/fmristat) for their work.

- Clustered m-sequence designs. These designs begin with an m-sequence design having maximal estimation efficiency and the tasks are permuted to improve the detection efficiency at the expense of estimation efficiency.
- Mixtures of the above types.

3.3 The influence of MRI physics on fMRI experiment design

An fMRI experiment must address a cognitive science question within the constraints of blocked and event-related experimental designs. In addition, the experimenter must select the appropriate parameters for the EPI sequence itself. These parameters are flip angle, echo time (T_E), repetition time (T_R), number of slices and slice thickness. There is also the option of choosing between various spin preparations prior to image acquisition including spin echo, gradient echo, inversion recovery, diffusion weighting, etc. For typical work at a main field strength of 1.5 T, a gradient echo sequence with a fat saturation pulse (see Section 2.1) is used. Other spin preparation options (e.g. spin echo, diffusion weighting, etc.) may be used for measuring BOLD signal from specific vascular sources at higher fields (see Section 4.2).

The selection of flip angle involves a trade-off with T_R in terms of signal intensity versus speed, but that trade-off is minor compared to other consequences of T_R selection (see below). So a selection of 90° is a good choice especially since isolated timing errors only affect one time point; i.e. with a 90° flip angle only one repetition is required for the spins to come to longitudinal equilibrium. This point is important for the beginning of a fMRI time-series acquisition. Most experiments collect five or so "dummy" volumes of data before the experiment proper commences. For a perfect 90° flip angle only one dummy volume is required to bring the spins to equilibrium but a few more are recommended, not only to compensate for imperfect flip angle control but also to allow the subject to get used to the noise made by the gradient coils during imaging.

The selection of T_E is a trade-off between having more signal for smaller T_E and more BOLD T_2^* contrast for longer T_E. Using fuzzy clustering methods (see Section 5.4) Barth *et al.* [28] show a steady increase in BOLD signal enhancement from short to long T_E for $T_E \in \{42 \text{ ms}, 70 \text{ ms}, 100 \text{ ms}, 130 \text{ ms}, 160 \text{ ms}\}$ except at $T_E = 130$ ms, where the enhancement was lower than for $T_E = 100$ ms and $T_E = 160$ ms. The lower enhancement at $T_E = 130$ ms is attributed to vascular and tissue spin dephasing (type-2 BOLD effect, see Section 4.2 and Fig. 4.3). At 4 and 7 T, Duong *et al.* [127] show that setting T_E to be approximately equal to brain gray matter T_2 leads to signal that is dominated by microvasculature signal over large vessel signal.

The selection of T_R also has two opposing mechanisms, a shorter T_R gives higher sampling of the HRF, while a longer T_R gives more NMR signal. Constable and Spencer [102] show, with both mathematical modeling and experimentation, that the statistical power gained by shortening T_R (from increased sampling of the HRF) far outweighs the NMR signal gain obtained by increasing T_R. So the experimenter should select the minimum T_R that will allow the collection of the required number of slices. Opposed to the criterion of selecting the shortest T_R possible is the observation that the false positive rate associated with uncorrected inferential statistics is reduced for longer T_R because the autocorrelation in the noise is reduced [365] (see Section 4.1.7).

The selection of the number of slices depends on the coverage desired. That coverage, in turn, is determined by the slice thickness. The trade-off in slice thickness selection is between increasing contrast with thinner slices and increasing SNR with thicker slices. Howseman *et al.* [227] show that higher SNR is more desirable because it leads to increased statistical significance in the activation maps. They show that acquired 1 mm thick slices smoothed to 8 mm yield less significant activation maps than maps made directly from acquired 8 mm thick slices and suggest that a slice thickness of 5–8 mm will optimize sensitivity and provide adequate localization. An exception to this recommendation is in the lower brain areas where the effect of signal drop out caused by susceptibility gradients from the sinuses can be countered with the use of thin slices. Howseman *et al.* further show that there is no effect on the significance of the map of slice order and in particular that there is no effect of T_1 spin tagging inflow in EPI-based fMRI experiments[†]. However, it should be noted that slice order can affect measurement of HRF latency and sequential slice order can lead to MRI signal cross-talk between the slices through overlapping slice profiles.

[†] Gao *et al.* [175] show more directly that the BOLD effect far outweighs blood inflow effects in EPI gradient echo based fMRI.

4

Univariate approaches: activation maps

The GLM approach to the analysis of neuroimages was first elucidated by Friston *et al.* [157] and the vast majority of fMRI data analysis techniques employed by neuroscientists use a GLM of one form or another. In this section we examine voxel-wise analysis of data; each voxel's time-series is subject to a statistical analysis and a summary statistic like Student's t or Fisher's F is computed. Color-coding the values of the statistic and plotting them as an image gives an SPM that is usually overlain on a high resolution anatomical image of the brain. Jernigan *et al.* [234] argue that it is important to retain the color-coding of an SPM to convey more information than just to show which voxels are above a cut off threshold. An alternative way to present an activation map is to present an amplitude (also color-coded) that is thresholded by the SPM values [107, 109], see Fig. 4.1.

4.1 The GLM – univariate approaches

A GLM of the time-series of an individual voxel in a 3D volume data set may be expressed as

$$\vec{y} = [X]\vec{\beta} + \vec{\epsilon}, \tag{4.1}$$

where $\vec{y}$ is the time-series data for one voxel, $\vec{\beta}$ is the parameter vector, $[X]$ is the design matrix and $\vec{\epsilon}$ is the residual or error vector. The columns of the design matrix characterize the parameters (explanatory variables) to be estimated. Consider, as an illustrative example, a blocked design with two volumes of images collected during the task followed by two volumes of images collected during rest at each epoch with three repeated epochs. (We assume that the volumes collected at the beginning of the time-series to allow the proton spins to come to longitudinal equalization have been discarded.) This simple design will result in the collection of 12 image volumes as a time-series. Consider the time-series from one 3D volume voxel, $\vec{y}$, in the simple data set; $\vec{y}$ is a 12-dimensional vector. Then the explicit GLM for our simple

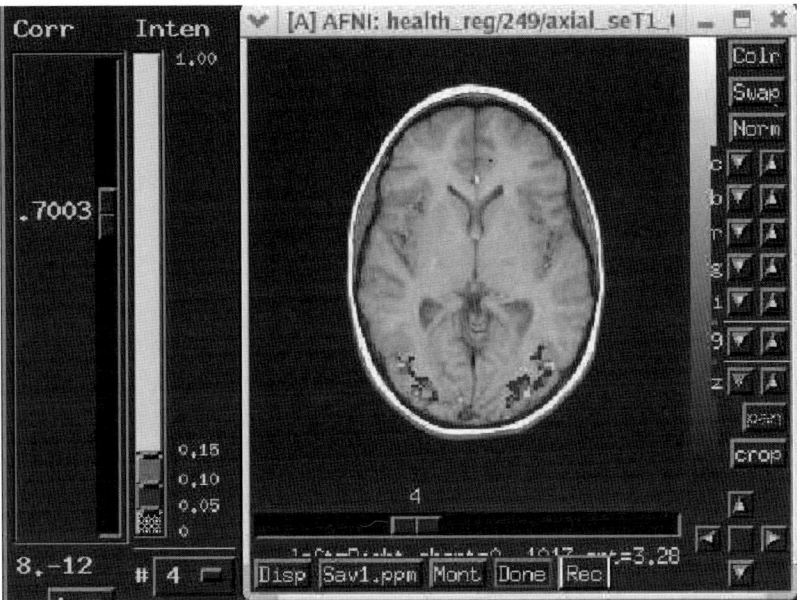

Fig. 4.1. An example of use of color-coding with an SPM. This image of the interface to the AFNI software shows a color-coded amplitude map thresholded with an SPM of correlation values. The correlation threshold in this example is set at 0.7003 by the slider on the interface and the color bar shows that the colors represent fractions of the maximum BOLD amplitude found in the data set. Many software packages for fMRI data processing have similar interfaces for displaying SPM related data. See also color plate.

experiment is

$$
\begin{bmatrix} y_1 \\ y_2 \\ y_3 \\ y_4 \\ y_5 \\ y_6 \\ y_7 \\ y_8 \\ y_9 \\ y_{10} \\ y_{11} \\ y_{12} \end{bmatrix}
=
\begin{bmatrix} 1 & 0 \\ 1 & 0 \\ 0 & 1 \\ 0 & 1 \\ 1 & 0 \\ 1 & 0 \\ 0 & 1 \\ 0 & 1 \\ 1 & 0 \\ 1 & 0 \\ 0 & 1 \\ 0 & 1 \end{bmatrix}
\begin{bmatrix} \beta_1 \\ \beta_2 \end{bmatrix}
+
\begin{bmatrix} \epsilon_1 \\ \epsilon_2 \\ \epsilon_3 \\ \epsilon_4 \\ \epsilon_5 \\ \epsilon_6 \\ \epsilon_7 \\ \epsilon_8 \\ \epsilon_9 \\ \epsilon_{10} \\ \epsilon_{11} \\ \epsilon_{12} \end{bmatrix},
\tag{4.2}
$$

where the two columns of $[X]$ are associated with the two indicator parameters β_1 (task) and β_2 (rest). In practice, additional columns might be added to account for drift and other factors (see Chapter 2). For event-related analysis, a single column could represent the HRF (this is also a better approach for blocked designs than the simple model of Equation (4.2)) or each column could represent a parameter describing an IRF. If the design matrix is of full rank (as is the case for our simple model), then an estimate of $\vec{\beta}$, $\hat{\vec{\beta}}$, may be found from the least squares solution

$$\hat{\vec{\beta}} = ([X]^T[X])^{-1}[X]^T\vec{y}. \tag{4.3}$$

Define $[X]^+$ to be a pseudoinverse of $[X]$. If $[X]$ is of full rank, then $[X]^+ = ([X]^T[X])^{-1}[X]^T$, otherwise, if $[X]$ is not of full rank (e.g. when the grand mean is modeled), then there exist a number of methods to compute $[X]^+$. With the pseudoinverse defined, Equation (4.3) may be written simply as

$$\hat{\vec{\beta}} = [X]^+\vec{y}. \tag{4.4}$$

The GLM is a multiple regression model and generalizes the idea of computing the correlation, as given by Equation (1.9), between a model HRF and the voxel time course [99]. When computing correlations there is the problem of what to choose for the model HRF. Popular choices are to let the model HRF be equal to an assumed IRF convolved with a step function defining the task paradigm (see Section 4.1.2). Another choice is to pick a representative time course given by the data from a region known to be active, as determined through visual inspection of the time course for example. Visual inspection of the time courses is easier for blocked designs where time courses that are correlated with the task presentation are more apparent (e.g. Fig. 1.1). Yet another choice, for blocked designs, is to average the response over the task epochs, for each voxel, and use the periodic extension of the average time course as the model HRF for the given voxel [389].

An effect of interest may be compared to a null hypothesis of no effect using a contrast vector, $\vec{c}$, to describe the effect of interest. Specifically the effect of interest is modeled by $\vec{c}^T\vec{\beta}$, and estimated by $\vec{c}^T\hat{\vec{\beta}}$. A t-test may then be used to test the null hypothesis that $\vec{c}^T\vec{\beta} = 0$ using the test statistic

$$t = \frac{\vec{c}^T\hat{\vec{\beta}}}{\sqrt{\text{var}(\vec{c}^T\hat{\vec{\beta}})}}, \tag{4.5}$$

where $\text{var}(\vec{c}^T\hat{\vec{\beta}})$ depends on the noise model adopted. For the simple design given by Equation (4.2) and a test to see if the signal during task is significantly different from that during rest, $\vec{c}^T = [-1\ 1]$. A useful noise model is that the noise is normally distributed as $N(0, \sigma^2[\Sigma])$. For the case in which noise between time points is not correlated, $[\Sigma] = [I]$ the identity matrix. With that $N(0, \sigma^2[\Sigma])$ noise model (assuming that $[\Sigma]$ is known),

$$\text{var}(\vec{c}^T\vec{\beta}) = \sigma^2[X]^+[\Sigma][X]^{+T}. \tag{4.6}$$

The variance σ^2 may be estimated from the residuals, $\vec{\epsilon}$, defined as

$$\vec{\epsilon} = \vec{y} - [X]\vec{\beta} = ([I] - [X][X]^+)\vec{y} = [R]\vec{y}, \tag{4.7}$$

where $[R] = [X]([X]^T[X])^+[X]^T$ is known as the residual matrix. With this residual matrix, the estimate, $\hat{\sigma}^2$ of σ^2 is

$$\hat{\sigma}^2 = \frac{\vec{\epsilon}^T\vec{\epsilon}}{\text{tr}[R]}, \tag{4.8}$$

where tr denotes trace and the effective degree of freedom, ν, to use with the t-test statistic of Equation (4.5) for comparison to the Student t distribution is

$$\nu = \frac{\text{tr}([R])^2}{\text{tr}([R]^2)}. \tag{4.9}$$

The map of t values given by Equation (4.5) is known as an SPM of t values or SPM$\{t\}$.

The design matrix and parameters usually need to be divided into effects of interest and effects of no interest (confounds) so that, with $\vec{\theta}$ representing the parameters of interest and $\vec{\phi}$ representing the confound parameters, Equation (4.1) becomes, at voxel i of n voxels,

$$\vec{y}_i = [A]\vec{\theta}_i + [B]\vec{\phi}_i + \vec{\epsilon}_i. \tag{4.10}$$

With a breakdown of the design matrix according to Equation (4.10), it becomes possible to characterize the response with the F statistic as discussed in Section 4.1.3. Otherwise, the use of contrasts and the t statistic as discussed above represent the main ideas behind the use of a GLM for characterizing and detecting BOLD response. We end this discussion with a review of a couple of methods that may potentially increase the power of the GLM approach but which are not in widespread use. Discussion of more conventional approaches resumes in Section 4.1.1.

The matrices $[A]$ ($p \times m$) and $[B]$ ($q \times m$) of Equation (4.10) are normally set a-priori but Ardekani et al. [16] show how both the dimension of and the bases (columns) for $[B]$ may be determined empirically if it is assumed that $[A]^T[B] = 0$ and $\vec{\epsilon}_i \sim N(0, \sigma^2[I])$. Defining the projection matrices $[P_A]$ and $[P_B]$ by

$$[P_A] = [A]([A]^T[A])^{-1}[A]^T \quad \text{and} \quad [P_B] = [B]([B]^T[B])^{-1}[B]^T, \tag{4.11}$$

Ardekani et al. show how the basis for $[B]$ may be chosen from the q eigenvectors of $([I] - [P_A])[R]$ having the largest eigenvalues, where

$$[R] = \frac{1}{n}\sum_{i=1}^{n}\vec{y}_i\vec{y}_i^T \tag{4.12}$$

is the covariance matrix of the data. The dimension q is found by minimizing the Akaike information criterion (AIC) given by

$$\text{AIC}(q) = -\ell([Y]; [\Theta], [\Phi], [B], \sigma^2) + K_q, \tag{4.13}$$

where ℓ is the log of the likelihood function given by

$$\ell([Y]; [\Theta], [\Phi], [B], \sigma^2) = -\frac{np}{2}\ln(2\pi\sigma^2) - \frac{1}{2\sigma^2}\sum_{i=1}^{n}(\vec{y}_i - [A]\vec{\theta}_i - [B]\vec{\phi}_i)^T$$

$$\times (\vec{y}_i - [A]\vec{\theta}_i - [B]\vec{\phi}_i) \tag{4.14}$$

and

$$K_q = nm + nq + [qp - qm - q(q+1)/2] + 1 \tag{4.15}$$

is the number of free parameters in the model. The maximum likelihood (ML) solution for the parameters (parameters that maximize ℓ) is given by the least squares solutions

$$\hat{\vec{\theta}}_i = ([A]^T[A])^{-1}[A]^T\vec{y}_i, \tag{4.16}$$

$$\hat{\vec{\phi}}_i = ([B]^T[B])^{-1}[B]^T\vec{y}_i \tag{4.17}$$

and the estimate for σ^2 is given by

$$\hat{\sigma}^2 = \frac{1}{np}\sum_{i=1}^{n}(\vec{y}_i - [A]\vec{\theta}_i - [B]\vec{\phi}_i)^T(\vec{y}_i - [A]\vec{\theta}_i - [B]\vec{\phi}_i). \tag{4.18}$$

Ardekani *et al.* further show that a constant false alarm rate (CFAR) matched subspace detector (MSD) for specifying the activation map is obtained by considering the F ratio

$$F(\vec{y}) = \frac{\vec{y}^T[P_A]\vec{y}/\sigma^2 m}{\vec{y}^T([I] - [P_A] - [P_B])\vec{y}/\sigma^2(p - q - m)} \tag{4.19}$$

and computing the activation map φ (with 1 as active) using the following uniformly most powerful (UMP) test of size α:

$$\varphi(F) = \begin{cases} 1 & \text{if } F > F_0, \\ 0 & \text{if } F \leq F_0, \end{cases} \tag{4.20}$$

where

$$\alpha = \int_{F_0}^{\infty} F_{m,p-q-m}\, dF \tag{4.21}$$

is the tail area, from F_0 on, of the F distribution with m numerator and $p - q - m$ denominator degrees of freedom.

The SNR for GLMs may be increased by "borrowing" signal from neighboring voxels. Let $\vec{z}_{i1}$ be the time-series data for a given voxel i and let $\vec{z}_{ij}$, $2 \leq j \leq L$ be the time-series of the voxels in a specified neighborhood of voxel i. Then we may consider the model of Equation (4.10) as being for the weighted average

$$\vec{y}_i = \sum_{j=1}^{L} w_j \vec{z}_{ij}. \tag{4.22}$$

Hossein-Zadeh *et al.* [226] show that the optimal weights $\vec{w}$ may be obtained by maximizing the objective function

$$\lambda(\vec{w}) = \frac{\vec{w}^T [Z]^T [P_A][Z]\vec{w}}{\vec{w}^T [Z]^T (I - [P_X])[Z]\vec{w}}, \tag{4.23}$$

where $[Z] = [\vec{z}_{i1} \ldots \vec{z}_{iL}]$ and $[P_X]$ is the projector into the column space of the design matrix defined analogously to Equations (4.11). Defining $[C] = [Z]^T [P_A][Z]$ and $[D] = [Z]^T (I - [P_X])[Z]$, $\vec{w}$ is the eigenvector corresponding to the largest eigenvalue of $[D]^{-1}[C]$. Once the weights are settled and a least squares solution to Equation (4.10) found and the resulting SPM thresholded to provide a provisional activation map, Hossein-Zadeh *et al.* apply a postprocessing procedure to trim the resulting active clusters (connected components of the activation map) of inactive voxels. The postprocessing proceeds in four steps: (1) For each provisionally active voxel, remove the nuisance effects from the time-series via $\vec{y} = \vec{z} - [P_B]\vec{z}$. (2) Find the "seed" time-series $\vec{r}$ in each connected cluster as the voxel having the most energy as defined by $\vec{y}^T [P_A]\vec{y}$. (3) For each voxel in each connected cluster compute

$$g = \frac{\vec{y}^T [P_A]\vec{y}}{\vec{r}^T [P_A]\vec{r}}. \tag{4.24}$$

(4) Declare all voxels for which $g < g_t$, where g_t is an empirically determined threshold, as inactive. Hossein-Zadeh *et al.* find that $g_t = 0.1$ works well.

4.1.1 Temporal smoothing

Based on the idea of filter matching, it has been shown that statistical power for detection may be increased by smoothing the fMRI voxel time-series in the temporal direction using a filter $[K]$, [151, 154, 455]. Typically Gaussian smoothing with $K_{ij} \propto \exp(-(i-j)^2/2\tau^2)$ is used, with $\tau^2 = 8$ s^2 [142]. With temporal smoothing, the GLM becomes

$$[K]\vec{y} = [K][X]\vec{\beta} + [K]\vec{\epsilon}, \tag{4.25}$$

and the estimates of $\vec{\beta}$ become

$$\hat{\vec{\beta}} = ([K][X])^+ [K]\vec{y}. \tag{4.26}$$

The process of multiplying the data and model by $[K]$ is known as coloring the data (see Section 4.1.7). The relevant variance to use in Equation (4.5) for $\hat{\vec{\beta}}$ of Equation (4.26), employing the simplification $[\Sigma] = [I]$ that the smoothing allows, is

$$\text{var}(\vec{c}^T \hat{\vec{\beta}}) = \vec{c}^T \, \text{var}(\hat{\vec{\beta}})\vec{c}, \tag{4.27}$$

where

$$\text{var}(\hat{\vec{\beta}}) = \sigma^2 ([K][X])^+ [V](([K][X])^+)^T \tag{4.28}$$

with $[V] = [K][K]^T$ and where

$$\hat{\sigma}^2 = \frac{\vec{e}^T \vec{e}}{\text{tr}([R][V])} \tag{4.29}$$

is an estimate for σ^2. In Equation (4.29) $\vec{e} = [R][K]\vec{y}$ and $[R]$ is the residual forming matrix. With temporal smoothing, the effective degrees of freedom are

$$\nu = \frac{\text{tr}([R][V])^2}{\text{tr}([R][V][R][V])}. \tag{4.30}$$

4.1.2 Modeling the impulse response function

The GLM design matrix may be constructed with an assumed IRF convolved with the presentation function (see Section 3.1) to give a model HRF. The model HRF is then used for BOLD signal detection. Or the design matrix may be designed with the goal of characterizing the IRF using a parametric model of the IRF. We review those two approaches to analysis here by looking at both kinds of IRF model.

One of the first models of the IRF assumed a Poisson form

$$h(\tau) = \lambda^\tau \, e^{-\lambda}/\tau!, \tag{4.31}$$

where τ and λ are positive integers [151]. Discrete convolution with the presentation paradigm then gives the model HRF (a column in $[X]$). Widely used refinements on Equation (4.31) include the gamma variate functions of Equations (3.4) and (4.195). An alternative way to model the HRF directly in a blocked design is to assume it has the functional form

$$f_k(t) = \sin(\pi t/(n+1)) \exp(-t/(nk)), \tag{4.32}$$

where n is the length of the task block in scans (number of task data volumes collected per epoch) with $k = 4$ to model an early response and $k = -1$ to model a late response [155]. Using the values of f_4 and f_{-1} at the scan times provides two columns for the design matrix $[X]$. Contrasts between the two columns can then reveal differences in response between different brain regions. Linear combinations of f_4 and f_{-1} are capable of modeling a variety of HRFs from standard unimodal responses to bimodal responses.

A GLM design with a single column representing the model HRF and a con-trast vector of $\vec{c} = [1]^T$ in Equation (4.5) is equivalent[†] to simply computing the correlation (see Equation (1.9)) between the time course and the model HRF through Equation (4.196). It is also possible to correlate the Fourier transform of the model HRF with the Fourier transform of the voxel time-series [26]. When using a correlation approach it is necessary to project out drift and other confounds as a preprocessing step (see Section 2.4.1).

Combining linear systems theory with the GLM allows one to construct a design matrix capable of characterizing the IRF h. Let u represent the stimulus function which, for event-related designs, is taken as

$$u(t) = \sum_{\text{events}} \delta(t - t_{\text{event}}), \tag{4.33}$$

then linear systems theory predicts a signal (HRF) y of the form

$$y(t) = u * h(t) + \epsilon(t), \tag{4.34}$$

where ϵ is the error time course. Let $\{g_b | 1 \le b \le B\}$ be a set of basis functions so that we may represent h as

$$h(t) - \sum_{b=1}^{B} \beta_b g_b(t), \tag{4.35}$$

then Equation (4.34) may be written as

$$y(t_n) = \sum_{b=1}^{B} u * g_b(t_n)\beta_b + \epsilon(t_n), \tag{4.36}$$

where t_n are the times sampled by the fMRI image time-series. Almost any set of basis functions can be used, including low order polynomials [60], with gamma variate functions (see Equations (3.4) and (4.195)) being common choices. Harms and Melcher [211] propose a set of five basis functions that they call the OSORU set, for onset, sustained, offset, ramp and undershoot, with each basis function capable of modeling those five physiologically observed components of the BOLD response (see Section 4.2 and Fig. 4.4). Clark [98] proposes orthogonal polynomials as a basis set. For blocked designs, a Fourier basis (sines and cosines) is particu-larly useful [15] and may be employed to model the HRF directly by skipping the

---

[†] It may be desirable to model the grand mean in $[X]$ to have a noise model with zero mean.

convolution with u given in Equations (4.34) and (4.36) (or equivalently setting u equal to Dirac's δ). Wavelet packet methods may be used to select the best wavelet packet basis from a local clustering point of view [317].

From Equation (4.36) we see that setting $X_{nb} = u*g_b(t_n)$ gives the design matrix. To adequately sample the IRF, nonconstant ISIs should be used [237]. As usual, columns may be added to the design matrix to represent effects of no interest like drift and physiological noise leaving a design matrix having B effects of interest.

As an alternative to modeling the IRF as a linear combination of basis functions, as given in Equation (4.35), one can model the amplitude of each point on a finite length IRF as a separate parameter by considering the discrete version of Equation (4.34) [44]. This leads to a finite impulse response (FIR) model of the form

$$y(t_n) = \sum_{m=1}^{M} u(t_{n-m+1})\beta_m, \qquad (4.37)$$

where M is the assumed finite length of the IRF and u is the stimulus function. In this case the design matrix entries are $X_{nm} = u(t_{n-m+1})$, not counting, of course, columns devoted to modeling effects of no interest. The approach of Equation (4.37) introduces less bias than the approach of Equation (4.36) into the estimation of the IRF when $B < M$, because no assumption is made about the shape of the IRF other than its length, and it generally provides a better estimate of the IRF shape [396].

4.1.3 Using the F statistic

To decide whether a voxel is activated or not from a design that models the IRF, an SPM of F statistics is appropriate. The appropriate F statistic may be determined using the extra sum of squares principle that works as follows. Partition the design matrix into columns of interest and no interest and the parameter vector into subvectors of interest and no interest and express the GLM as

$$\vec{y} = [X_H | X_r] \begin{bmatrix} \vec{\beta}_1 \\ - \\ \vec{\beta}_r \end{bmatrix} + \vec{\epsilon}, \qquad (4.38)$$

where the subscript H refers to effects of interest and r to effects of no interest. Let $S_R = \vec{\epsilon}^T \vec{\epsilon}$ be the residual sum of squares for the full model and $S_r = \vec{\epsilon}_r^T \vec{\epsilon}_r$ be the residual sum of squares for the reduced model

$$\vec{y} = [X_r]\vec{\beta}_r + \vec{\epsilon}_r \qquad (4.39)$$

and let $p = \text{rank}[X]$, $p_r = \text{rank}[X_r]$ and $q = p - p_r$. Let $S_H = S_r - S_R$ and assume the the noise model for $\vec{\epsilon}$ is $N(0, \sigma^2[I])$, where $[I]$ is the identity matrix (i.e. independent (white) Gaussian noise), then the test statistic[†]

$$F = \frac{S_H/v_0}{S_R/v} \qquad (4.40)$$

may be compared to $F_{v_0,v}$ to test the null hypothesis of no activation, where the degrees of freedom of the numerator $v_0 = q$ and the denominator $v = N - p$, where $N = \dim \vec{y}$. Subsequently an SPM{F} may be plotted. If we define the projectors

$$[R] = [I] - [X][X]^+, \qquad (4.41)$$

$$[R_r] = [I] - [X_r][X_r]^+, \qquad (4.42)$$

$$[M] = [R] - [R_r], \qquad (4.43)$$

then Equation (4.40) can be rewritten as

$$F = \frac{\vec{y}^T [M] \vec{y}/\mathrm{tr}[M]}{\vec{y}^T [R] \vec{y}/\mathrm{tr}[R]}. \qquad (4.44)$$

If the noise model is $N(0, \sigma^2[\Sigma])$, then the numerator of Equation (4.44) will no longer follow a $\chi^2_{v_0}$ distribution under the null hypothesis. Instead it will follow a linear combination of χ^2_1 distributions that, by the Satterthwaite approximation, can be approximated by a gamma distribution. Under that approximation, the appropriate F statistic to compute is [455]

$$F = \frac{\vec{y}^T [M] \vec{y}/\mathrm{tr}([M][\Sigma])}{\vec{y}^T [R] \vec{y}/\mathrm{tr}([R][\Sigma])}, \qquad (4.45)$$

which may be compared to $F_{v_0,v}$, where

$$v_0 = \frac{\mathrm{tr}([M][\Sigma])^2}{\mathrm{tr}([M][\Sigma][M][\Sigma])} \quad \text{and} \quad v = \frac{\mathrm{tr}([R][\Sigma])^2}{\mathrm{tr}([R][\Sigma][R][\Sigma])} \qquad (4.46)$$

(see Equation (4.30)[‡]). To apply Equation (4.45), knowledge of $[\Sigma]$ is required. One approach to estimating $[\Sigma]$ is to compute a pooled estimate over all the voxels (the SPM2 software does that). However, it is known that the time-series autocorrelation varies over the brain, due to the varying physiological processes that cause the autocorrelation (see Section 2.4.2), so that another approach is to model the autocorrelation function ρ (see Equation (1.5)) for each voxel [259]. Assume, in the following discussion, for simplicity, that normalized time is used such that $t_n = n$ for the data point at position n in the time-series. When ρ is given, the entries of the estimated autocorrelation matrix $[\hat{\Sigma}]$ are given by

$$\hat{\Sigma}_{ij} = \rho(|i - j|). \qquad (4.47)$$

[†] The terms S_H and S_R are variously known as the sum of squares due to the model and sum of squares due to the error respectively or as the "between" sum of squares and "within" sum of squares respectively from an ANOVA/ANCOVA point of view.

[‡] The theory developed for the F statistic may be applied to the t statistic under the equivalence $t_v^2 = F_{1,v}$.

The autocorrelation function, ρ, is found by fitting a model to the estimated autocorrelation function

$$\hat{\rho}(k) = \frac{\sum_{i=1}^{N-k}(\vec{\epsilon}_i - \bar{\epsilon})(\vec{\epsilon}_{i+k} - \bar{\epsilon})}{\sum_{i=1}^{N-k}(\vec{\epsilon}_i - \bar{\epsilon})^2}, \tag{4.48}$$

where $\bar{\epsilon}$ is the mean of the residual vector components. Kruggel *et al.* [259] find that the "damped oscillator" model

$$\rho(k) = \exp(a_1 k)\cos(a_2 k) \tag{4.49}$$

fits fMRI data well. The standard AR(1) model may be obtained from Equation (4.49) by setting $a_2 = 0$.

Kiebel *et al.* [249] argue, using restricted maximum likelihood (ReML) estimators of σ^2 for the noise model $N(0, \sigma^2[\Sigma])$, that the F statistic of Equation (4.45) should be used with $\nu = \text{tr}[R]$ instead of the formula given in Equation (4.46). Then they extend their work to the case where $\vec{\epsilon} \sim N(0, [C_\epsilon])$, where $[C_\epsilon] = \sum_{i=1}^{m}\lambda_i[Q_i]$ is described by multiple variance parameters λ_i that may be estimated using iterative ReML methods. In that case, they recommend the use of

$$F = \frac{\vec{y}^T[M]\vec{y}/\text{tr}([M][C_\epsilon])}{\vec{y}^T[R]\vec{y}/\text{tr}([R][C_\epsilon])} \tag{4.50}$$

with

$$\nu_0 = \frac{\text{tr}([M][\hat{C}_\epsilon])^2}{\text{tr}([M][\hat{C}_\epsilon][M][\hat{C}_\epsilon])} \tag{4.51}$$

and

$$\nu = \frac{\sum_i \hat{\lambda}_i^2 \, \text{tr}([M][Q_i])^2}{\sum_i \hat{\lambda}_i^2 \, \text{tr}([M][Q_i])^2 \, \text{tr}([\hat{R}_{\text{ML}}][Q_i][\hat{R}_{\text{ML}}][Q_i])^{-1}}, \tag{4.52}$$

where

$$[\hat{R}_{\text{ML}}] = [I] - [X]([X]^T[\hat{C}_\epsilon]^{-1}[X])^{-1}[X]^T[\hat{C}_\epsilon]^{-1}. \tag{4.53}$$

If $[\hat{C}_\epsilon]$ is available, then another alternative is to prewhiten the data by multiplying the model $\vec{y} = [X]\vec{\beta} + \vec{\epsilon}$ by $[C_\epsilon]^{-1/2}$ to obtain a model with independent errors [67] (see Section 4.1.7).

4.1.4 Nonlinear modeling: Volterra kernels

By running multiple experiments to measure the HRF from widely spaced single trials of varying duration (SD, see Section 3.1), Glover [184] is able to show significant deviation from expected linear system behavior. The extent of non-linearity has also been observed to vary across the cortex, becoming nonlinear for stimuli of less than 10 s in the primary auditory cortex, less than 7 s in the primary motor cortex and less than 3 s in the primary visual cortex [320, 407]. Shorter duration stimuli lead to a more nonlinear response than longer duration stimuli [281, 429] and using an ISI shorter than 2 s also leads to nonlinear response [64, 113]. The BOLD response appears to be more nonlinear closer to the neuronal source than farther away [360]. The nonlinearity of the BOLD signal has been experimentally shown to occur because of the dependence of the signal on multiple physiological parameters, specifically on cerebral blood volume (CBV), cerebral blood flow (CBF) and deoxyhemoglobin concentration [177] (see Section 4.2).

Nonlinear aspects of the HRF may be accounted for in a GLM by using Volterra kernels [162]. To do this, generalize the linear approach of Equation (4.34) to the completely general[†] nonlinear form[‡]

$$
\begin{aligned}
y(t) = h^0 &+ \int_{-\infty}^{\infty} h^1(\tau_1) u(t - \tau_1) \, d\tau_1 \\
&+ \int_{-\infty}^{\infty} \int_{-\infty}^{\infty} h^2(\tau_1, \tau_2) u(t - \tau_1) u(t - \tau_2) \, d\tau_1 \, d\tau_2 \\
&+ \int_{-\infty}^{\infty} \int_{-\infty}^{\infty} \int_{-\infty}^{\infty} h^3(\tau_1, \tau_2, \tau_3) u(t - \tau_1) u(t - \tau_2) u(t - \tau_3) \, d\tau_1 \, d\tau_2 \, d\tau_3 \\
&+ \text{etc.} + \epsilon(t),
\end{aligned}
\tag{4.54}
$$

where h^n is the nth order Volterra kernel. The linear systems model includes only the zeroth and first order kernels. To apply Equation (4.54) we truncate it to second order and assume a causal (so that the integrals start at 0) model with finite memory (so that the integrals stop at T) to arrive at

$$
\begin{aligned}
y(t) = h^0 &+ \int_0^T h^1(\tau_1) u(t - \tau_1) \, d\tau_1 \\
&+ \int_0^T \int_0^T h^2(\tau_1, \tau_2) u(t - \tau_1) u(t - \tau_2) \, d\tau_1 \, d\tau_2 \\
&+ \epsilon(t).
\end{aligned}
\tag{4.55}
$$

[†] At least for analytic functions.
[‡] In practice, Equation (4.34) will have an overall mean term similar to h^0 of Equation (4.54).

With a set of basis functions $\{g_b | 1 \leq b \leq B\}$, the Volterra kernels may be written as

$$h^1(\tau_1) = \sum_{i=1}^{B} c_b^1 g_i(\tau_1), \tag{4.56}$$

$$h^2(\tau_1, \tau_2) = \sum_{j=1}^{B} \sum_{i=1}^{B} c_{ij}^2 g_i(\tau_1) g_j(\tau_2). \tag{4.57}$$

If we let

$$x_i(t) = \int_{-\infty}^{\infty} g_i(\tau_1) u(t - \tau_1) \, d\tau_1, \tag{4.58}$$

then Equation (4.55) becomes

$$y(t) = h^0 + \sum_{i=1}^{B} c_i^1 x_i(t) + \sum_{j=1}^{B} \sum_{i=1}^{B} c_{ij}^2 x_i(t) x_j(t) + \epsilon(t). \tag{4.59}$$

Equation (4.59) may be seen to be a GLM by setting

$$\vec{\beta} = \begin{bmatrix} h^0 & c_1^1 & \cdots & c_B^1 & c_{1,1}^2 & \cdots & c_{B,B}^2 \end{bmatrix}^T \tag{4.60}$$

and the first column of $[X]$ to be 1s, the next B columns to be

$$[x_i(t_1) \cdots x_i(t_N)]^T,$$

where N is the number of time points measured, $1 \leq i \leq B$, and the final B^2 columns to be

$$[x_i(t_1) x_j(t_1) \cdots x_i(t_N) x_j(t_N)]^T$$

for $1 \leq i, j \leq B$. As usual, effects of no interest may be added as additional columns and detection achieved via use of an F statistic. The model of Equation (4.54) may be simplified by assuming that the Volterra kernels are tensor products, e.g. $h^2(\tau_1, \tau_2) = h^1(\tau_1) h^1(\tau_2)$, in which case it may be shown that the Volterra series is a McLaurin expansion of

$$y(t) = f\left(\int_{-\infty}^{\infty} h^1(\tau_1) u(t - \tau_1) \, d\tau_1 \right), \tag{4.61}$$

where f is a nonlinear function.

An alternative approach to nonlinear modeling with Volterra kernels, when stochastic inputs are used, is to use Wiener kernels [391]. In terms of Wiener kernels, k^i, Equation (4.55) is equivalent to

$$y(t) = [k^0] + \left[\int_0^T k^1(\tau_1)u(t - \tau_1) \, d\tau_1 \right]$$

$$+ \left[\int_0^T \int_0^T k^2(\tau_1, \tau_2)u(t - \tau_1)u(t - \tau_2) \, d\tau_1 \, d\tau_2 - \sigma^2 \int_0^T h^2(\tau, \tau) \, d\tau \right]$$

$$+ \epsilon(t) \tag{4.62}$$

when $u(t)$ is white Gaussian noise input with variance σ^2 [241]. The terms in square brackets in Equation (4.62) are the Wiener functionals. The Wiener kernels are then the expected values

$$k^0 = \langle y(t) \rangle, \tag{4.63}$$

$$k^1(\tau_1) = \langle y(t) \, u(t - \tau_1) \rangle, \tag{4.64}$$

$$k^2(\tau_1, \tau_2) = \frac{1}{2!} \langle y(t) \, u(t - \tau_1) \, u(t - \tau_2) \rangle \quad \text{for} \tau_1 \neq \tau_2. \tag{4.65}$$

In practice, better estimates of the Wiener kernels are obtained by specifying an m-sequence input [40] (see Section 3.2) instead of white Gaussian noise input for u.

4.1.5 Multiple tasks

In all the cases discussed so far we may separate more than one process within one time-series [351]. This is done by using several task functions u_i with models of the form

$$y(t) = \sum_{i=1}^{N} u_i * h(t) + \epsilon(t) \tag{4.66}$$

or the nonlinear equivalent where N tasks are assumed. In that case, the number of columns of interest (not counting the mean) in the design matrix is multiplied by N. It is also possible to design graded stimuli [54] so that u of Equation (4.34) or (4.66) would be of the form

$$u(t) = \sum_j a_j \delta(t - t_j) \tag{4.67}$$

with a_j not necessarily equal to 1.

4.1.6 Efficiency of GLM designs

The efficiency of a given GLM design is given in [310] as

$$\text{efficiency} = \frac{1}{\sigma^2 \vec{c}^T ([X]^T [X])^{-1} \vec{c}} \tag{4.68}$$

or if a contrast matrix $[c]$ is specified

$$\text{efficiency} \propto \frac{1}{\text{tr}(\sigma^2 [c]^T ([X]^T [X])^{-1} [c])}. \tag{4.69}$$

By formulating a model for the expected value $\langle [X]^T [X] \rangle$ for use in Equation (4.68) for stochastic designs, where the occurrence of stimuli at fixed times is given by a probability, Friston *et al.* [164] show that:

- Long SOA designs are less efficient than rapid presentation designs.
- Nonstationary designs, where the probability of occurrence varies in time, are more efficient than stationary designs.
- The most efficient design is the blocked design.
- With multiple trial types in the same time-series, the efficiency depends on the type of contrasts desired. For example, the efficiency of detecting differences grows as the probability for event occurrence increases but the probability of detecting evoked responses declines as the probability of event occurrence increases.

Following Liu *et al.* [282] a clean distinction between estimation efficiency and detection efficiency may be made by explicitly breaking the GLM down into the effect of interest, the model of the IRF, and effects of no interest (nuisance effects) as follows:

$$\vec{y} = [X]\vec{\beta}_H + [S]\vec{\beta}_r + \vec{\epsilon}, \tag{4.70}$$

where $[X]\vec{\beta}_H$ models the effect of interest and $[S]\vec{\beta}_r$ models the nuisance effects. With that breakdown, the estimate of the IRF is given by (assuming matrices of full rank for simplicity)

$$\vec{\beta}_H = ([X]^T [P_S^\perp][X])^{-1} [X]^T [P_S^\perp][X]\vec{y}, \tag{4.71}$$

where $[P_S^\perp] = [I] - [S]([S]^T [S])^{-1}[S]^T$ is a projection matrix that projects $\vec{y}$ into the subspace perpendicular to the span of the columns of $[S]$. Using $[X_\perp] = [P_S^\perp][X]$ we can rewrite Equation (4.71) as

$$\vec{\beta}_H = ([X_\perp]^T [X_\perp])^{-1} [X_\perp]^T \vec{y}. \tag{4.72}$$

Then, following the rationale that the efficiency is the inverse of the variance of the estimated parameters that lead also to Equations (4.68) and (4.69), we may define *estimation* efficiency as

$$\xi = \frac{1}{\sigma^2 \, \text{tr}[([X_\perp]^T [X_\perp])^{-1}]}. \tag{4.73}$$

Detection is based on using the F statistic to decide between

$$H_0: \quad \vec{y} = [S]\vec{\beta}_r + \vec{\epsilon} \tag{4.74}$$

and

$$H_1: \qquad \vec{y} = [X]\vec{\beta}_H + [S]\vec{\beta}_b + \vec{\epsilon}. \qquad (4.75)$$

Liu *et al.* [282] show that *detection* power may be usefully quantified by the Rayleigh quotient

$$R = \frac{\vec{\beta}_H^T [X_\perp]^T [X_\perp]\vec{\beta}_H}{\vec{\beta}_H^T \vec{\beta}_H}. \qquad (4.76)$$

The quantities ξ and R are opposed to each other: as designs are changed to increase R (moving toward blocked designs), then ξ must necessarily decrease and vice versa. Another factor in fMRI design is the perceived randomness of task presentation to the experimental participant. Randomness can be required for reducing confounds due to anticipation and habituation. Liu *et al.* [280, 283] define an entropy measure, H_r to quantify the randomness of an fMRI experimental design from the cognitive point of view. They find that H_r and ξ tend to increase together.

4.1.7 Noise models

Many specific (e.g. heartbeat and respiration) and nonspecific physiological and physical processes involved in the generation of an fMRI BOLD signal produce a time-series with correlated noise. Failure to account for the correlated noise in the analysis of the signal can result in a loss of statistical power. Preprocessing to remove the systematic non-BOLD variation is possible according to the methods outlined in Section 2.4 and these methods tend to reduce the correlations in the noise but more general analysis methods can estimate and reduce the correlations further. Three main approaches along these lines are to correct the relevant degrees of freedom (variance correction) associated with the computed statistics (see Section 4.1.3), to color the data and model so as to end up with noise terms with small relative autocorrelation (see Section 4.1.1), or prewhiten the data and model to remove the autocorrelation from the noise terms (see the end of Section 4.1.3). The smoothing approach outlined in Section 4.1.1 is a coloring approach where multiplication by $[K]$ introduces a known autocorrelation that is essentially accounted for in Equations (4.29) and (4.30). (Note that $[K]$ may include a low pass filter component in addition to the smoothing aspect to provide a band-pass filter.) Following Woolrich *et al.* [449] we can describe these approaches in more detail.

The basic GLM is[†]

$$\vec{y} = [X]\vec{\beta} + \vec{e} \qquad (4.77)$$

with $\vec{e}$ assumed to be $N(0, \sigma^2[V])$. Then there exists a square, nonsingular matrix $[K]$ such that[‡] $[V] = [K][K]^T$ and $\vec{e} = [K]\vec{\epsilon}$, where $\vec{\epsilon} \sim N(0, \sigma^2[I])$. Next introduce

---

[†] Note the slight change of symbols used here from the previous subsections.

[‡] The matrix $[K]$ is known as the square root of $[V]$.

a filter matrix $[S]$ by which we can color the data and model of Equation (4.77) to produce

$$[S]\vec{y} = [S][X]\vec{\beta} + \vec{\eta}, \tag{4.78}$$

where $\vec{\eta} \sim N(0, \sigma^2[S][V][S]^T)$. From Equation (4.78) the ordinary least squares (OLS) estimate of $\vec{\beta}$ is

$$\hat{\vec{\beta}} = ([S][X])^+[S]\vec{y} \tag{4.79}$$

(see Equation (4.26)). The variance of a contrast of these parameter estimates is

$$\mathrm{var}\left(\vec{c}^T\hat{\vec{\beta}}\right) = k_{\mathrm{eff}}\sigma^2, \tag{4.80}$$

where

$$k_{\mathrm{eff}} = \vec{c}^T([S][X])^+[S][V][S]^T(([S][X])^+)^T\vec{c} \tag{4.81}$$

(see Equation (4.28) where $[V]$, in the notation of Equation (4.81), is set to $[I]$). The noise variance parameter σ^2 may be estimated using

$$\hat{\sigma}^2 = \frac{\vec{\eta}^T\vec{\eta}}{\mathrm{tr}([R][S][V][S]^T)} \tag{4.82}$$

(see Equation (4.29)), where $[R] = [I] - [S][X]([S][X])^+$ is the residual forming matrix with the residuals given by

$$\vec{r} = [R][S]\vec{y} = [R]\vec{\eta}. \tag{4.83}$$

With this setup we can classify the three approaches to dealing with noise autocorrelation mentioned above by:

- *Coloring.* Where $[S]$ is a filter that is designed so that, relative to the filter, we can assume $[V] = [I]$. This is the approach outlined in Section 4.1.1.
- *Variance correction.* Where $[S] = [I]$ (i.e. no filter) is used and an estimate of $[V]$ is used to compute

$$k_{\mathrm{eff}} = \vec{c}^T[X]^+[V]([X]^+)^T\vec{c}. \tag{4.84}$$

  This is the approach outlined in Section 4.1.3.
- *Prewhitening.* Assuming $[V]$ can be determined, set $[S] = [K]^{-1}$ so that

$$k_{\mathrm{eff}} = \vec{c}^T([X]^T[V]^{-1}[X])^{-1}\vec{c}. \tag{4.85}$$

  This approach gives the best linear unbiased estimates (BLUE), or Gauss-Markov estimates, of $\vec{\beta}$. Methods of estimating $[V]$ are reviewed below. (The damped oscillator model of Kruggel *et al.* [259] is reviewed in Section 4.1.3.)

The usual approach to estimating $[V]$ or its transformed version $[S][V][S]^T$ is to estimate the autocorrelation matrix of the residuals $\vec{r}$ (Equation (4.83)) which turns out to be $[R][S][V][S]^T[R]^T$ and not $[S][V][S]^T$. In practice, because of the typical frequency spectrum structure of fMRI data, the difference in the two correlation matrices is small [449].

One of the earlier approaches to modeling the autocorrelation of $\vec{r}$ used $1/f$ (where f = frequency) type models[†] where the spectral density of $\vec{r}$, $\vec{a}$, was fit to one of the following two three-parameter models[‡]

$$|a(\omega)| = \frac{1}{k_1(\omega/2\pi + k_2)} + k_3 \qquad (4.86)$$

or a decaying exponential model of the form

$$|a(\omega)| = k_1 e^{-\omega/(2\pi k_2)} + k_3, \qquad (4.87)$$

where k_1, k_2 and k_3 are the parameters to be fit [469]. A spatially smoothed version of the above approach was also tried [2] and it was found that MRI time-series of water phantoms also had the same $\sim 1/f$ noise structure indicating that $1/f$ processes are due to physical MRI processes and not physiological processes.

Woodrich *et al.* [449] further try tapers such as the Tukey window given by

$$\hat{\rho}(\tau) = \begin{cases} \frac{1}{2}(1 + \cos[\pi\tau/M]) \, r(\tau) & \text{if } \tau < M, \\ 0 & \text{if } \tau \geq M, \end{cases} \qquad (4.88)$$

where $M \sim 2\sqrt{N}$, with N being the length of the time-series, is close to optimal. They also try a nonparametric pool adjacent violators algorithm (PAVA), multi-tapering and AR parametric model estimation. An AR process of order p, denoted AR(p), is given by

$$r(t) = \phi_1 r(t-1) + \phi_2 r(t-2) + \cdots + \phi_p r(t-p) + e(t), \qquad (4.89)$$

where e is a white noise process and $\{\phi_i\}$ are the parameters. The SPM software uses AR(1) models based on averaged fits to residuals over the whole brain to refine the estimate of a global ϕ_1 [459]. Spatial smoothing can be used to improve the results for voxel-wise noise modeling approaches and Worsley *et al.* [459] introduce an additional procedure for reducing the bias of the noise model estimate.

In prewhitening approaches, it is necessary to have an estimate of $[V]$ to construct the filter. A two-step process usually suffices to estimate $[V]$ [449]. First set $[S] = [I]$ in Equation (4.78), find the first estimate $[V_1]$ from the residual and construct

---

[†] These are more properly referred to as "modified $1/f$ models" because the amplitude of the spectral density is assumed to follow $1/f$. Conventional $1/f$ models, having fractal scaling properties, assume that the power (amplitude squared) follows $1/f$ [166].

[‡] Note the transition from the vector $\vec{a}$ to function a.

$[S] = [V_1]^{-1/2}$. With this $[S]$ solve Equation (4.78) again and estimate $[V_2]$ from the resulting residuals. Use $[S] = [V_2]^{-1/2}$ as the prewhitening matrix.

Friston *et al.* [166] propose that different filtering strategies (selection of $[S]$) may be classified according to *validity*, *robustness* and *efficiency*. Validity refers to the accuracy of the computed p values, efficiency refers to minimizing the variance of the estimated parameters and robustness refers to the sensitivity that the computed p values have to violations of the assumed model (especially estimates of the autocorrelation matrix $[V]$). Friston *et al.* quantify efficiency and bias to show how robustness may be traded off with efficiency. To illustrate these quantities, introduce $[K_i]$ and $[K_a]$ to represent the inferred and the actual convolution matrix $[K]$ respectively so that $[V_i] = [K_i][K_i]^T$ and $[V_a] = [K_a][K_a]^T$. Efficiency is defined as

$$\text{efficiency} = \frac{1}{\sigma^2 \vec{c}^T ([S][X])^+ [S][V_i][S]^T (([S][X])^+)^T \vec{c}} \tag{4.90}$$

(see Equation (4.68)). Bias is defined as

$$\text{bias}([S],[V_i]) = 1 - \frac{\text{tr}([R][S][V_i][S]^T) \vec{c}^T ([S][X])^+ [S][V_a][S]^T (([S][X])^+)^T \vec{c}}{\text{tr}([R][S][V_a][S]^T) \vec{c}^T ([S][X])^+ [S][V_i][S]^T (([S][X])^+)^T \vec{c}}. \tag{4.91}$$

Negative bias leads to p values that are too small (too lax), positive bias leads to p values that are too large (too conservative). Prewhitening is known to produce a minimum variance (maximum efficiency) filter but Friston *et al.* show that a band-pass filter for $[S]$ produces a minimum bias filter and therefore provides a more robust procedure for estimating $\vec{\beta}$ and associated p values.

The conclusion that band-pass filtering is optimal from a bias point of view is dependent on the use of an a-priori assumed distribution of the test statistic under the null hypothesis. When the H_0 distribution is determined empirically, prewhitening produces less bias [69]. The null distribution of a test statistic may be determined empirically by permuting the data and computing the resultant test statistic [341]. When the noise is serially correlated, the data permutation must be done carefully so that the noise associated with the permuted data has the same autocorrelation properties as the original data. Two approaches that preserve noise autocorrelation have been used to permute fMRI time-series data: a Fourier method and a wavelet method. With the Fourier method, the (complex-valued) Fourier transformation of the time-series data for each voxel is computed, the phases are shuffled and the inverse Fourier transform computed to yield the "permuted" data set [264]. The wavelet method is based on the fact that $1/f$ processes have fractal properties that are preserved under the right wavelet transform [69]. Specifically, under an orthogonal wavelet transform, the expected correlation between two wavelet coefficients $w_{j,k}$

and $w_{j',k'}$ for a fractional Brownian motion or other $1/f$-like noise process is[†]

$$E[r(w_{j,k}, w_{j',k'})] \sim \mathcal{O}(|2^j k - 2^{j'} k'|^{2(H-R)}), \qquad (4.92)$$

where R is the number of vanishing moments of the wavelet function and H is the Hurst exponent[‡] of the noise process. If $R > 2H + 1$, it can be shown that the wavelet coefficient correlations will decay hyperbolically fast within levels and exponentially fast between levels. The Hurst coefficient for fMRI noise processes can be argued to be $H < 1$ so R of at least 4 is required (i.e. Daubechies's compactly supported wavelets with four vanishing moments should be used [114]). The use of wavelets with higher R does not achieve more decorrelation because compactly supported orthogonal wavelets with more vanishing moments have larger support, leading to a higher risk of computational artifact. Similar reasoning [69] leads to the recommendation that the maximum scale level, J, to which the wavelet decomposition should be computed is J such that $N/2^{J-1} \geq 2R$. Data shuffling in the wavelet domain using the appropriate wavelet transform then proceeds as follows. Compute the wavelet transform, shuffle the wavelet coefficients within each scale level and, compute the inverse wavelet transform to end up with "permuted" data that have the same noise characteristics as the original data. Computing the test statistic for different wavelet permuted data will give an empirical null distribution of the test statistic. The use of a translation-invariant wavelet transform to compute permutation statistics produces an overcomplete set of wavelet coefficients, a subset of which must be chosen for permutation and reconstruction purposes [225].

Kamba *et al* [239] model the source of noise directly, using a measurement of the global signal (see Section 2.4.3) in a model of the BOLD response. Their model is an autoregressive model with exogenous inputs (ARX model) that may be expressed as

$$A(q) y(t) = \vec{B}(q)^T \vec{u}(t - k) + \epsilon(t), \qquad (4.93)$$

where the first component of $\vec{u}$, u_1, represents the stimulus input and the second component, u_2, represents the global signal input. The functions A and $\vec{B}$ are polynomials of the shift operator q:

$$A(q) = 1 + a_1 q^{-1} + \cdots + a_{n_a} q^{n_a}, \qquad (4.94)$$

$$B_i(q) = 1 + b_{i,1} q^{-1} + \cdots + b_{i,n_b} q^{n_b}, \quad i = 1, 2. \qquad (4.95)$$

The magnitudes of $b_{1,j}$ characterize the activations.

[†] The $\mathcal{O}$ is Landau's big oh notation meaning "of the same order as".
[‡] See Section 5.3 for more information about and applications of the Hurst exponent.

4.1.8 Type I error rates

An activation map, or more directly an SPM, consists of $\sim 4 \times 10^3$ univariate statistics, one for each voxel. At $\alpha = 0.05$ we can therefore expect ~ 2000 false positives. A correction to α (or to the reported p value) is necessary to correct for multiple comparisons to correctly give the probability that a voxel declared as active is actually a false positive. A straightforward approach is to use Bonferroni correction and multiply the per voxel p by the number of voxels, v, in the brain. If there is spatial correlation in the data, the Bonferroni correction will be too conservative and activations will be missed [453]. An MRI image that is Fourier reconstructed at the same matrix size as the acquisition matrix will have a point spread function (PSF) whose FWHM is comparable to the pixel size [387] so there will be very little spatial correlation in the plane of the image in such a case. In the slice direction there will be some spatial correlation due to overlapping slice profiles (partial volume effect). Otherwise spatial correlation is largely introduced by data preprocessing. The interpolation involved in image realignment (see Section 2.3) will induce spatial correlation as will deliberately imposed spatial smoothing designed to increase statistical power through averaging.

Even in the absence of spatial correlation, the statistical power of Bonferroni for large numbers of comparisons is low, so alternative p value correction methods are needed in general. One approach is to control the false discovery rate more directly. To test v voxels using the false discovery rate method [180], let q be the rate of false discovery that the researcher is willing to tolerate and let $p_{(1)}, p_{(2)}, \ldots, p_{(v)}$ be the uncorrected p values associated with each voxel in order from smallest to largest p value. Let r be the largest i for which

$$p_{(i)} \leq \frac{i}{q} \frac{q}{c(v)}, \tag{4.96}$$

where, for the case of nonnegative spatial correlation (includes spatial independence),

$$c(v) = \sum_{i-1}^{v} 1/i \approx \ln v + \gamma, \tag{4.97}$$

where $\gamma \approx 0.577$ is Euler's constant. Declare the voxels with i such that $1 \leq i \leq r$ to be active.

Forman *et al.* [141] describe a method to correct p value based on false discovery rates in clusters using the idea that larger clusters are less likely than smaller clusters to appear by chance. Genovese *et al.* [179] define a test-retest reliability, based on experimental repetition of a test study, that may then be applied to the study of interest.

When the data are spatially correlated, the theory of Gaussian random fields (GRF) [1] may be used to provide corrected p values. Let the set of voxels that represent the brain be denoted by S and the set of active voxels, as defined by a thresholding procedure (all voxels whose test statistic is above a threshold t), be denoted by A_t. Using techniques from algebraic topology, the Euler characteristic, χ_t, of the set A_t may be defined. For a sufficiently high threshold t, the component subsets of A_t will have no holes (be homotopic to zero[†]) and $\chi_t = m_t$, where m_t is the number of maxima of the SPM of test statistics in S that are greater than t. To apply GRF theory, the SPM must be of Z statistics. So an SPM of t statistics (a t-field, see Equation (4.5)) is transformed into a Gaussianized t-field (Gt-f) on the basis of equivalent p values by setting, for each voxel,

$$Z = \Phi^{-1}(\Psi_\nu(t)), \tag{4.98}$$

where Φ is the standard normal cumulative density function (CDF) and Ψ_ν is the CDF of the Student's t distribution with ν degrees of freedom. With the Gt-f and GRF theory we can define three types of probability, or p values [246]:

(i) The probability, under H_0, that a *voxel* at a maximum of the Z-field has a value, Z_{max}, that is greater than a given threshold t is approximately equal to $E(m_t)$, the expected value of $\chi_t = m_t$ [150, 454]. That is

$$p(Z_{max} > t) \approx E(m_t) = V\, ||[\Lambda]||^{1/2}\, (2\pi)^{-(D+1)/2}\, \mathrm{He}_D(t)\, e^{-t^2/2}, \tag{4.99}$$

where V is the volume of the brain (search) set S, $||[\Lambda]||$ is the determinant of $[\Lambda]$, D is the dimension of the data set (2 or 3 depending on whether the SPM is of a slice or volume) and He_D is the Hermite polynomial of degree D ($\mathrm{He}_2(t) = t^2 - 1$, $\mathrm{He}_3(t) = t^3 - 3t$). The matrix $[\Lambda]$ is the variance–covariance matrix of the partial derivatives of the spatial correlation process in the D spatial directions. If X represents that process, then in (x, y, z) 3D space,

$$[\Lambda] = \begin{bmatrix} \mathrm{var}\left(\dfrac{\partial X}{\partial x}\right) & \mathrm{cov}\left(\dfrac{\partial X}{\partial x}, \dfrac{\partial X}{\partial y}\right) & \mathrm{cov}\left(\dfrac{\partial X}{\partial x}, \dfrac{\partial X}{\partial z}\right) \\[2ex] \mathrm{cov}\left(\dfrac{\partial X}{\partial x}, \dfrac{\partial X}{\partial y}\right) & \mathrm{var}\left(\dfrac{\partial X}{\partial y}\right) & \mathrm{cov}\left(\dfrac{\partial X}{\partial y}, \dfrac{\partial X}{\partial z}\right) \\[2ex] \mathrm{cov}\left(\dfrac{\partial X}{\partial x}, \dfrac{\partial X}{\partial z}\right) & \mathrm{cov}\left(\dfrac{\partial X}{\partial y}, \dfrac{\partial X}{\partial z}\right) & \mathrm{var}\left(\dfrac{\partial X}{\partial z}\right) \end{bmatrix}. \tag{4.100}$$

The matrix $[\Lambda]$ needs to be estimated from the data (see below). The SPM software denotes the probabilities given by Equation (4.99) as voxel level p values in its output.

[†] Homotopies are families of continuous functions that may be continuously transformed into each other. To be homotopic to zero means that any function that is homotopic to the surface of a sphere (in 3D – homotopic to a circle in 2D) can be continuously transformed, within the given set, to a point (zero). That is, the set has no holes that the shrinking homotopy can get hung up on.

(ii) The probability, under H_0, that the size n_t of a *cluster* in a GRF above a high threshold t exceeds k voxels is [152]

$$p(n_t > k) = 1 - \exp(E(m_t)\, e^{-\beta k^{2/D}}), \qquad (4.101)$$

where $E(m_t)$ represents the expected number of clusters above the threshold t, as described above, and

$$\beta = \left(\frac{\Gamma(D/2 + 1)E(m_t)}{V\Phi(-t)} \right)^{2/D} \qquad (4.102)$$

with V and Φ as defined above. The SPM software denotes the probabilities given by Equation (4.101) as cluster level p values in its output.

(iii) The probability, under H_0, that the number of clusters in a *set of clusters* in a GRF above threshold t and of size greater than n is greater than k is given by [158]

$$p(C_{n,t} > k) = 1 - \sum_{i=0}^{k-1} \Upsilon(i, E(m_t)p(n_t > n)), \qquad (4.103)$$

where $C_{n,t}$ is the number of clusters of size n above threshold t and

$$\Upsilon(i, p) = \frac{p^i\, e^{-p}}{i!}. \qquad (4.104)$$

The SPM software denotes the probabilities of Equation (4.103) as set level p values in its output.

The above probabilities may also be estimated empirically using permutation methods [214, 215] (see Section 4.1.7). Sensitivity and regional specificity both diminish as one moves from the voxel level (Equation (4.99)) to cluster level (Equation(4.101)) to set level (Equation (4.103)). It is also possible to give a p value that represents the probability of finding a given cluster size, n_t and peak height, $Z_{\max}$ simultaneously under H_0 [362]. If the FWHM of the PSF is not constant in all directions (as may happen with smoothed maps mapped to a 2D cortical surface – see Section 5.1.1) or the correlations are not stationary, then a correction to ordinary GRF theory is required [457]. All GRF methods require the estimation of $[\Lambda]$ which may be done using a (vector-valued) map of the residuals from the GLM fit [246, 361]. To estimate $[\Lambda]$, first form the standardized residuals $\vec{S}$ from the residuals $\vec{R}$ by

$$\vec{S} = \frac{\vec{R}}{\sqrt{\vec{R}^T \vec{R}}}, \qquad (4.105)$$

then the components of $[\hat{\Lambda}]$, $\hat{\lambda}_{ii}$ (the covariances can be assumed to be zero), can be computed from

$$\hat{\lambda}_{ii} = \lambda_v \frac{v - 2}{(v - 1)n} \sum_{x \in S} \sum_{j=1}^{n} \left(\frac{\partial S_j(x)}{\partial l_i} \right)^2, \qquad (4.106)$$

where n is the length of the fMRI time-series, S is the search region, l_i represent the three directions (for 3D GRFs), ν is the effective degrees of freedom (see for example, Equation (4.30)), the partial derivatives are estimated using finite difference techniques and λ_ν is a correction factor given by

$$\lambda_\nu = \int_{-\infty}^{\infty} \frac{(t^2 + n - 1)^2}{(\nu - 1)(\nu - 2)} \frac{\psi_\nu(t)^3}{p(t)^2} \, dt, \tag{4.107}$$

where ψ_ν is the probability density function (PDF) of a t distribution with ν degrees of freedom, t is the threshold and $p(t) = \phi(\Phi^{-1}(1 - \Psi_\nu(t)))$ with ϕ being the PDF of the standard normal distribution.

The methods covered in this section and in Sections 4.1.3 and 4.1.7 all assume that the random variables represented by the columns of the design matrix have a Gaussian distribution (i.e. the population being sampled is normal). Hanson and Bly [210] show that, in fact, the BOLD-parameter population is better described by a gamma distribution than by a normal distribution. This observation is why, they speculate, many investigators need to threshold their activation maps at very low p value thresholds to avoid excessive type I errors that are visible from visual inspection of the maps (e.g. the observation of activation in nongray matter). If the computed p values were based on the assumption of a population gamma distribution, the thresholds used would correspond to higher p values.

4.1.9 Combining and comparing activation maps between subjects

Thus far we have reviewed methods for computing activation maps for individuals. Experimental tests of cognitive neuroscience hypotheses require testing multiple subjects under multiple conditions. So analysis at the group level is necessary. In principle, it is possible to analyze group data altogether with one giant GLM using, for example, columns in the design matrix to represent indicator variables for different conditions. Such an approach is prohibitively expensive from the computational point of view and not necessary. It is possible to test any hypothesis that can be formulated under a single-level GLM with a hierarchical analysis in which "summary statistics" produced at the individual level are analyzed in a second GLM that reflects the cognitive conditions more directly [38].

The simplest group analysis one might want to do is to produce an "average" or summary of the activation maps from the group. Lazar *et al.* [270, 307] give several methods that can be used for combining t-statistics of voxels (see Equation (4.5)) from k individuals:

- *Fisher's method* [140] in which the group statistic is

$$T_F = -2 \sum_{i=1}^{k} \ln p_i, \tag{4.108}$$

where p_i is the p value associated with the t-statistic from subject i. Under H_0, T_F follows χ^2_{2k}.

- *Tippett's method* [422] in which the group statistic is

$$T_T = \min_{i=1}^{k} p_i, \qquad (4.109)$$

where p_i is the p value associated with the t-statistic from subject i. The null hypothesis is rejected if $T_T < 1 - (1 - \alpha)^{1/k}$ when the acceptable type I error rate is set at α. A generalization of this procedure is to consider the rth smallest p value and reject H_0 if that p value is smaller than a constant that depends on k, r and α [442].

- *Stouffer's method* [410] in which the group statistic is

$$T_S = \sum_{i=1}^{k} \frac{\Phi^{-1}(1 - p_i)}{\sqrt{k}}, \qquad (4.110)$$

where Φ is the standard normal CDF. Stouffer's statistic is compared to the standard normal distribution.

- *Mudholkar and George's method* [329] in which the group statistic is

$$T_M = -\sqrt{\frac{3(5k + 4)}{k\pi^2(5k + 2)}} \sum_{i=1}^{k} \ln\left(\frac{p_i}{1 - p_i}\right) \qquad (4.111)$$

which is compared to t_{5k+4}.

- *Worsley and Friston's method* [458] where

$$T_W = \max_{i=1}^{k} p_i \qquad (4.112)$$

rejecting H_0 if $T_W < \alpha^{1/k}$.

- *Average t method* in which the group statistic is

$$T_A = \sum_{i=1}^{k} \frac{t_i}{\sqrt{k}}, \qquad (4.113)$$

where t_i is the t-statistic from subject i. The statistic T_A is an approximation to T_S and is considered to follow the standard normal distribution under H_0.

Of all these methods, Tippett's method produces the worst results because no averaging is done [270], see Fig. 4.2. Methods for combining estimated parameters (e.g. $\vec{c}^T \vec{\beta}$) directly, in mixed-effects models, are discussed below. McNamee *et al.* [307], using jack-knife (leave-one-out) analysis find that Fisher's method is the most sensitive to individual outlier behavior, a mixed-effects model is the most robust and Stouffer's method is in between[†].

[†] Lazar *et al.* [270, 307] use the FIASCO software package to do their work. FIASCO can be found at http://www.stat.cmu.edu/~fiasco/.

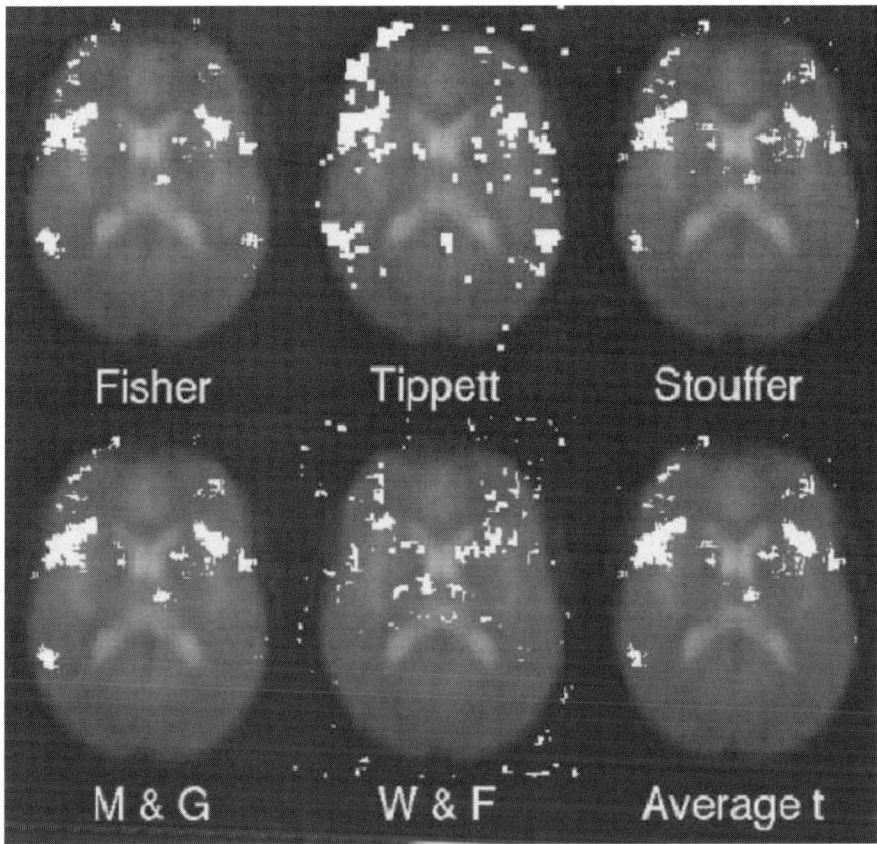

Fig. 4.2. Comparison of group SPM averaging methods, taken from [270] (used with permission). M & G = Mudholkar and George's method, W & F = Worsley and Friston's method. Tippett's method produced the most false positives, Worsley and Friston's method produced the fewest true positives. Data are presented in Talairach coordinates and the task was a memory-guided visual saccade task.

Friston *et al.* [163] use conjunction analysis to make inferences about the proportion of the population that would show an effect (an activation in a given region) based on the observation that all n subjects in a study showed the effect. If $\alpha_{\min}$ represents the uncorrected p value obtained by setting T_W of Equation (4.112) equal to $\alpha_{\min}$, then the theory of n intersecting GRFs may be used to compute a corrected p value, p_n. If $1 - \alpha_c$ is the specificity required, then the minimum population proportion γ_c that would show activation in the regions for cases where all n subjects showed activation is

$$\gamma_c \geq \frac{((\alpha_c - p_n)/(1 - p_n))^{1/n} - \alpha_{\min}}{1 - \alpha_{\min}}. \tag{4.114}$$

To account for the variation of individual brain organization, the relevant p value, p_i, for use in Equation (4.112) can be the minimum p value from a predefined ROI.

A fixed-effects model combines the parameter of interest y_i in each subject to obtain an average θ for a group of k subjects according to the model

$$y_i = \theta + \epsilon_i, \tag{4.115}$$

where $E(\epsilon_i) = 0$ and $\text{var}(\epsilon_i) = V_i = 1/w_i$. With this model the parameter estimate is the weighted average

$$\hat{\theta} = \frac{\sum_{i=1}^{k} w_i y_i}{\sum_{i=1}^{k} w_i} \tag{4.116}$$

and, using

$$\text{var}(\hat{\theta}) = \frac{1}{\sum_{i=1}^{k} 1/V_i}, \tag{4.117}$$

the associated test statistic is the usual t-statistic

$$T_X = \frac{\hat{\theta}}{\sqrt{\text{var}(\hat{\theta})}}. \tag{4.118}$$

In the case that all V_i are equal, $T_X = T_S$, where T_S is Stouffer's statistic.

In case we do not believe that there is a fixed population θ and, instead, believe that the variation in y_i is due in part to real differences between subjects and not just measurement error, we use a mixed-effects model (also known as a random-effects model by some) of the following form

$$y_i = \theta_i + \epsilon_i, \tag{4.119}$$

$$\theta_i = \theta + e_i, \tag{4.120}$$

where $E(\epsilon_i) = 0$ and $\text{var}(\epsilon_i) = V_i$, as with the fixed-effects model, and $E(e_i) = 0$ and $\text{var}(e_i) = \sigma_\theta^2$. The estimate of θ in this model is

$$\hat{\theta}^* = \frac{\sum_{i=1}^{k} w_i^* y_i}{\sum_{i=1}^{k} w_i^*}, \tag{4.121}$$

where $w_i^* = 1/(s_i^2 + \hat{\sigma}_\theta^2)$ with s_i^2 being an estimate of V_i. The test statistic is

$$T_R = \frac{\hat{\theta}^*}{\sqrt{\text{var}(\hat{\theta}^*)}}, \tag{4.122}$$

where

$$\text{var}(\hat{\theta}^*) = \frac{1}{\sum_{i=1}^{k} 1/(V_i + \sigma_\theta^2)}, \tag{4.123}$$

a quantity that will be larger than the corresponding fixed-effects value given by Equation (4.117). The difficulty with mixed-effects models is the estimation of $\hat{\sigma}_\theta^2$. One approach is

$$\hat{\sigma}_\theta^2 = s^2 - \frac{\sum_{i=1}^{k} s_i^2}{k}, \tag{4.124}$$

where s^2 is the sample variance of $\{y_i | 1 \le i \le k\}$. A drawback to the use of Equation (4.124) is that it may give negative results. Bayesian methods for estimating random-effects variance components are reviewed in Section 4.4.

The extension of fixed- and mixed-effects models of simple averaging to more general models, like the popular ANOVA models used by cognitive neuroscientists, is relatively straightforward. It is informative to view these more general models in their place in a two-level hierarchical model. Following Beckmann *et al.* [38], we have at the first level, our usual GLM for subject k

$$\vec{y}_k = [X_k]\vec{\beta}_k + \vec{\epsilon}_k, \tag{4.125}$$

where $E(\vec{\epsilon}_k) = 0$, $\text{cov}(\vec{\epsilon}_k) = [V_i]$. Let

$$\vec{y} = \left[\vec{y}_1^T \, \vec{y}_2^T \cdots \vec{y}_N^T \right]^T, \tag{4.126}$$

$$[X] = \begin{bmatrix} [X_1] & [0] & \cdots & [0] \\ [0] & [X_2] & \cdots & [0] \\ \vdots & & \ddots & \vdots \\ [0] & \cdots & [0] & [X_N] \end{bmatrix}, \tag{4.127}$$

$$\vec{\beta} = \left[\vec{\beta}_1^T \, \vec{\beta}_2^T \cdots \vec{\beta}_N^T \right]^T, \tag{4.128}$$

$$\vec{\epsilon} = \left[\vec{\epsilon}_1^T \, \vec{\epsilon}_2^T \cdots \vec{\epsilon}_N^T \right]^T, \tag{4.129}$$

then the two-level model is

$$\vec{y} = [X]\vec{\beta} + \vec{\epsilon}, \tag{4.130}$$

$$\vec{\beta} = [X_G]\vec{\beta}_G + \vec{\eta}, \tag{4.131}$$

where $[X_G]$ is the group-level design matrix (e.g. an ANOVA), $\vec{\beta}_G$ represents the group-level parameters and $\vec{\eta}$ is the group-level error vector with $E(\vec{\eta}) = 0$, $\text{cov}(\vec{\eta}) = [V_G]$ and $\text{cov}(\vec{\epsilon}) = [V]$, a block-diagonal form of the $[V_k]$ matrices. Equations (4.130) and (4.131) generalize Equations (4.119) and (4.120). Composing Equations (4.130) and (4.131) gives the large single-level model mentioned at the beginning of this section:

$$\vec{y} = [X][X_G]\vec{\beta}_G + \vec{\gamma}, \tag{4.132}$$

where $\vec{\gamma} = [X]\vec{\eta} + \vec{\epsilon}$, $E(\vec{\gamma}) = 0$ and $\text{cov}(\vec{\gamma}) = [X][V_G][X]^T + [V]$. In general, the estimate of $\vec{\beta}_G$ obtained from using the BLUE estimate of $\vec{\beta}$ from Equation (4.130) on the left hand side of Equation (4.131) will not be the same as the BLUE estimate of $\vec{\beta}_G$ obtained from the direct least squares solution of Equation (4.132). To ensure that the estimate of $\vec{\beta}_G$ from the two-level model coincides with the estimate from the single level model we must define

$$[V_{G2}] = [V_G] + ([X]^T[V]^{-1}[X])^{-1} \tag{4.133}$$

and estimate $\vec{\beta}_G$ using

$$\hat{\vec{\beta}}_G = ([X_G]^T[V_{G2}]^{-1}[X_{G2}])^{-1}[X_G]^T[V_{G2}]^{-1}\hat{\vec{\beta}}, \tag{4.134}$$

$$\text{cov}(\hat{\vec{\beta}}_G) = ([X_G]^T[V_{G2}]^{-1}[X_{G2}])^{-1}. \tag{4.135}$$

If it is assumed that $[V_G] = [0]$ in Equation (4.133), then the two-level model is a fixed-effects model. Otherwise, it is a mixed-effects model and sophisticated (iterative Bayesian) procedures may be required to estimate $[V_G]$ if heteroscedasticity (unequal within subjects variance) is assumed. When homoscedasticity is assumed, i.e. that $[V_G] = \sigma_\theta^2[I]$, closed form, least squares solutions are available for $\hat{\vec{\beta}}$. Beckmann *et al.* [38] give examples for some of the more popular second-level GLMs like paired and unpaired t-tests and ANOVA F-tests. The FSL software[†] may be used to implement these models.

As with the first level analysis, an empirical determination of the H_0 distribution of the test statistic at the second level is also possible. In that case permuting the data is more straightforward than the methods outlined in Section 2.4 because there is generally no correlation in noise between subjects. An application of permutation tests for mixed plot ANOVA designs in fMRI is given in [413].

4.2 Modeling the physiology of the hemodynamic response

Many BOLD detection methods rely on a model of the BOLD response over time, as we have seen in previous sections. Those HRF models were and are largely empirical, based on observed responses. To validate and improve the IRF/HRF models used for the analysis of fMRI time-series it is necessary to understand the physiology of the BOLD response. In this section we look at work aimed at understanding that physiology.

The fMRI signal is based on the BOLD contrast mechanism whose basic physiological features were elucidated soon after it was discovered [347, 349, 350, 395]. Following Ugurbil *et al.* [426], we outline those basic features here.

[†] Software available at http://www.fmrib.ox.ac.uk/fsl/.

BOLD contrast is based on the fact that deoxyhemoglobin is paramagnetic. A *magnetic field intensity*, H, can cause *magnetization intensity*, M, to occur in material objects. The amount of magnetization is determined by the *susceptibility* χ as given by $M = \chi H$. Here H (and later B) refer to the magnitudes of the vector fields. Note that the *magnetic field induction*, B, is related to magnetic field intensity by permeability μ defined by $B = \mu H$. But variation in μ is small compared to variations in χ in different tissues. Materials may be classified according to their susceptibility, roughly:

1. *Diamagnetic.* $\chi < 0$: these materials repel magnetic fields (reduce the intensity). A rod-shaped piece of diamagnetic material will line up perpendicular to the magnetic field. Most body tissues are diamagnetic.
2. *Paramagnetic.* $\chi > 0$: these materials attract (enhance) magnetic fields. A rod-shaped piece of paramagnetic material will line up parallel to the magnetic field.
3. *Ferromagnetic.* $\chi \gg 0$ (typically $\chi > 1000$): these materials are composed of magnetized clusters of atomic magnets that easily line up with the external magnetic field. They stick to magnets.

A "susceptibility gradient" will surround the paramagnetic deoxyhemoglobin molecule. That is, the magnetic field near the deoxyhemoglobin will be higher than it will be further away. So the Larmor frequency of the proton spins (usually in water) will be higher near the deoxyhemoglobin than farther away. Suppose we have an infinitely long blood vessel at an angle θ to the magnetic field. If Y represents the fraction of oxygenated blood present, then it can be shown that the change in Larmor frequency from ω_0 is given by

$$\Delta\omega_B^{in} = 2\pi \, \Delta\chi_0 \, (1 - Y) \, \omega_0 \left(\cos^2 \theta - \frac{1}{3} \right) \qquad (4.136)$$

inside the cylinder and

$$\Delta\omega_B^{out} = 2\pi \, \Delta\chi_0 \, (1 - Y) \, \omega_0 \, (r_b/r)^2 \, \sin^2 \theta \, \cos 2\phi \qquad (4.137)$$

outside the cylinder, where $\Delta\chi_0$ is the maximum susceptibility difference expected in the presence of fully deoxygenated blood. The effect of the $\Delta\omega_B$ is to cause spins within a voxel to dephase, thus causing T_2^* signal decay.

The BOLD phenomenon is dependent on the diffusion of the signal-producing water molecules. The diffusive motion can lead to "irreversible T_2^*". The effect of diffusion can be divided into two regimes: dynamic and static averaging.

Dynamic averaging occurs when the diffusive path length during T_E is long compared to the susceptibility gradient, $\partial\Delta\omega_B/\partial r$. This occurs, for example, in blood vessels around red blood cells or around capillaries or venuoles. (Note that there is no BOLD from fully oxygenated arterial blood.) Dynamic averaging produces irreversible T_2^*.

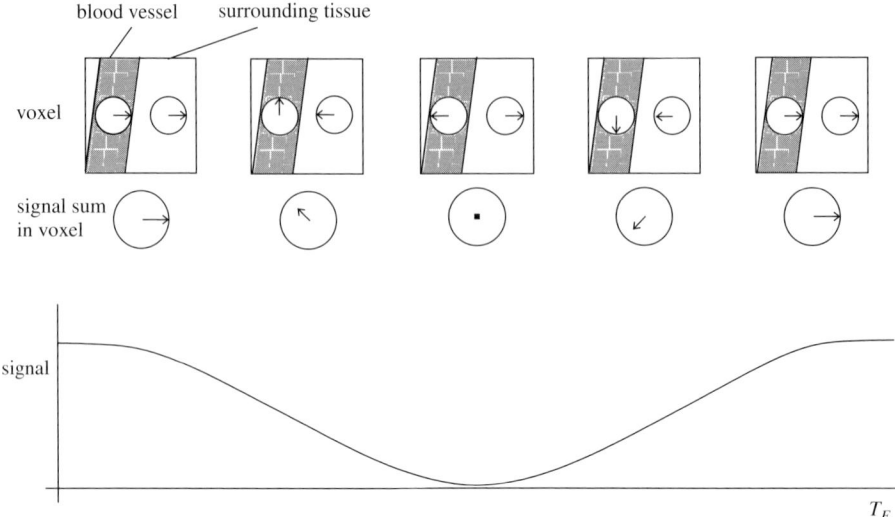

Fig. 4.3. Type 2 BOLD effect. The boxes represent a single voxel. The phase of the signal from the blood vessel, S_i, in that voxel will be in or out of phase with the signal from the tissue, S_e, as T_E is changed because of the different precession rates between blood and tissue signal. In addition, the overall signal will decay with longer T_E because of T_2 decay of the two components. (See Equation (4.142) for a signal model that incorporates S_i and S_e.) The BOLD effect will manifest itself during activation with a signal from the vascular component that is more coherent with the extravascular component.

Static averaging occurs when the diffusive path length is small compared to the susceptibility gradient, which happens around large blood vessels. Static averaging produces reversible T_2^*. Signal loss due to reversible T_2^* may be recovered as usual with the 180° RF pulse of a spin echo sequence.

The definition of what constitutes a large blood vessel increases with increasing field strength, B_0, because the susceptibility gradient, represented by $\Delta\omega_B$, is a function of $\omega_0 = \gamma B_0$. At higher field strengths, static averaging should diminish and be replaced by more sensitive dynamic averaging around smaller blood vessels.

A third type of BOLD signal is from the so-called "type 2" mechanism (Fig. 4.3). In this case, the blood vessels and surrounding tissue occupy a similar volume within a voxel. The bulk Larmor frequencies of the protons in the blood will be different from the bulk Larmor frequencies in the tissue. So the signal from the blood and the surrounding tissue will go out of and into phase and lose coherence as they both decay according to different T_2 values. Type 2 BOLD signal loss[†] is reversible

---

[†] Signal loss in this sense happens during rest; during activation, the source of dephasing (deoxyhemoglobin) is reduced resulting in less loss and the BOLD signal enhancement.

(recoverable) with a spin echo sequence. Note that type 2 signal has, by definition, an intravascular component and an extravascular component; the sum of these components is essentially modeled in Equation (4.142).

BOLD signal may come from water molecules in two sources: intravascular and extravascular. And we have seen that there are reversible and irreversible T_2^* phenomenon associated with each source:

- Extravascular:
  - reversible T_2^* around large blood vessels;
  - irreversible T_2^* around small blood vessels.
- Intravascular:
  - reversible T_2^* caused by type-2 BOLD effect;
  - irreversible T_2^* caused by water molecules moving in and out of red blood cells.

Separating the sources has been accomplished experimentally through the use of diffusion weighted MRI, where the addition of bipolar diffusion gradients to a sequence attenuates the signal from diffusing water. These same gradients will also kill signal from macroscopically flowing water. Therefore, a "diffusion sequence" can be used to eliminate intravascular BOLD signal [274]. In a similar vein, spin echo sequences may be used to eliminate the BOLD signal from reversible T_2^* effects[†].

Using a diffusion sequence at 1.5 T kills all the BOLD signal, which shows that the BOLD signal at 1.5 T must be intravascular. At higher fields, there is still some BOLD signal from a diffusion experiment and thus some extravascular signal. At very high fields (~9 T) a diffusion sequence does not affect the BOLD signal at all, implying that at very high fields all signal is extravascular.

Using a spin echo sequence for BOLD at 1.5 T results in a large reduction in signal over gradient echo sequences showing that most of the BOLD signal at 1.5 T is type 2 BOLD. This result is consistent with an investigation by Hoogenraad *et al.* [222], who use inversion recovery sequences to suppress signal from gray matter, white matter or cerebral spinal fluid (CSF) to find that activated voxels almost always come from a gray matter or CSF voxel that contains a vein. So the BOLD "spatial resolution" at 1.5 T must be comparable to the voxel size because the type 2 effect is intrapixel – although the small signal seen using spin echo sequences could be intravascular signal from large blood vessels. At 4 T it has been shown that BOLD signal can be localized to 700 μm [315]. A study by Kim *et al.* [252], however, finds that, while the BOLD response correlates linearly with the underlying neural activity, localization of that neural activity to only supramillimeter scales is possible.

[†] The use of a magnetization transfer pulse in the EPI sequence, for saturating signal from macromolecules, has the potential to also identify tissue types associated with the BOLD signal [472].

The BOLD signal, being dependent on the amount of deoxyhemoglobin present, depends on three factors:

- CBV – cerebral blood volume,
- CBF – cerebral blood flow,
- CMRO$_2$ – cerebral metabolic rate of O$_2$ consumption.

The CBV component[†] is, of course, dependent on changes in CBF. A model for the BOLD signal change ΔS as a function of CBF and CMRO$_2$ is

$$\frac{\Delta S}{S} = A \left[1 - \left(\frac{\text{CBF}}{\text{CBF}_0} \right)^{\alpha-\beta} \left(\frac{\text{CMRO}_2}{\text{CMRO}_{2,0}} \right)^{\beta} \right], \tag{4.138}$$

where the subscripts 0 refer to baseline values. Using the assumed values $A = 0.25$, $\alpha = 0.38$ and $\beta = 1.5$, Uludağ *et al.* [427] plot CBF/CBF$_0$ versus CMRO$_2$/CMRO$_{2,0}$ to find a value for the coupling n between CBF and CMRO$_2$, defined as

$$n = \frac{\Delta \text{CBF}}{\text{CBF}_0} \bigg/ \frac{\Delta \text{CMRO}_2}{\text{CMRO}_{2,0}}, \tag{4.139}$$

to be fairly tight at a value of $n = 2.2 \pm 0.15$. Uludağ *et al.* are also able to use a model for oxygen extraction fraction, E, given by

$$E = E_0 \left(\frac{\text{CMRO}_2}{\text{CMRO}_{2,0}} \bigg/ \frac{\text{CBF}}{\text{CBF}_0} \right) \tag{4.140}$$

and an assumed baseline of $E_0 = 40\%$ to find $E = 33 \pm 1.4\%$ during activation, $46 \pm 1.4\%$ during an eyes-closed rest period and $43 \pm 1.2\%$ during an eyes-open rest period. Schwarzbauer and Heinke [394] use perfusion sensitive spin tagging techniques with conventional BOLD fMRI to measure $\Delta \text{CMRO}_2 = 4.4 \pm 1.1\%$.

   The important point to remember is that the *less* deoxyhemoglobin present (the more the oxygenated blood), then the *more* the BOLD signal. The presence of deoxyhemoglobin causes susceptibility gradients which decreases T_2^* (because of spin dephasing). And, at a fixed T_E, the smaller (or shorter) T_2^* is, the less MRI signal there will be. So the higher the ratio CBF/CMRO$_2$, the higher the BOLD signal. At the beginning of the BOLD response, it is believed that the CMRO$_2$ increases at first but that an increase in CBF shortly following that is responsible for most of the BOLD signal. The basic BOLD response (IRF) appears as shown in Fig. 4.4. There is an initial negative dip, followed by an increase and decline,

---

[†] Many experimental investigations of the relationship between CBV, CBF and CMRO$_2$ measure changes in total CBV with changes in CBF (e.g. [250] where the FAIR MRI spin tagging technique [251] is used to measure CBF or [242] where an exogenous relaxation enhancing contrast agent is used to measure CBV). Lee *et al.* [275] find that a 100% increase in CBF leads to ~31% increase in total CBV of which only half (15%) of the increase is due to venous CBV change, so they suggest that many experimental investigations that measure CBV need to make a correction to arrive at a venous CBV value.

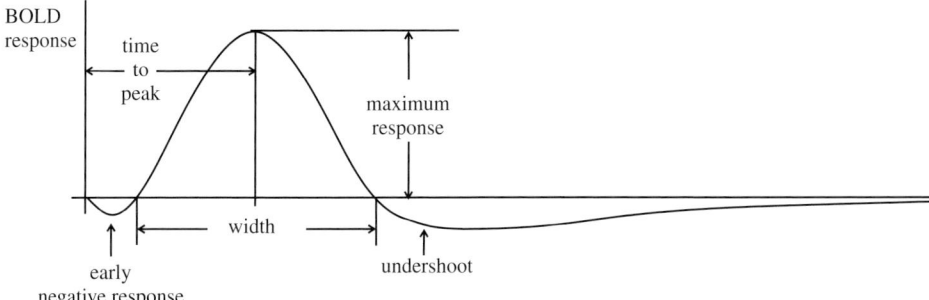

Fig. 4.4. A typical BOLD response to a short stimulus. The major features of the basic BOLD function are well modeled by the bivariate gamma function of Equation (4.195). The initial dip will occur within a delay period of ~2 s before the rise begins. The rise to maximum takes ~6–10 s followed by a decline that takes a similar amount of time [76]. The undershoot may be due to elevated oxygen extraction after the blood flow has returned to baseline.

finishing off with an undershoot and returning to baseline. These major features are well modeled by the bivariate gamma function of Equation (4.195). The initial dip has received a considerable amount of attention from some investigators because it represents an opportunity for increased spatial localization of the underlying neuronal signal [77, 126]. This increased localization is possible because the initial dip presumably represents an increased demand for oxygen by the neurons before the hemodynamic system has had a chance to respond with increased CBF.

In some regions of the brain, in some tasks, a response that is the negative of that shown in Fig. 4.4 is sometimes seen. Three theoretical mechanisms have been proposed to explain the negative BOLD response (NBR) [435]:

1. The mechanism is *vascular steal*, where oxygenated blood is taken from the NBR regions to supply the regions showing a positive BOLD response.
2. The mechanism is *active neuronal suppression*, where neural oxygen usage is reduced below baseline.
3. The mechanism is an *extended initial dip*, in which the CBF does not increase over the oxygen consumption in a neuronally active region.

Experimental work by Shmuel *et al.* [399] supports the active neuronal suppression hypothesis.

Logothetis *et al.* [284, 285] report that the BOLD response is due to local field potentials (LFPs). Logothetis *et al.* measure both multi unit activity (MUA) and LFP while doing an fMRI BOLD measurement on monkeys. According to them, MUA corresponds to the output of a neural population; LFP corresponds mostly to a weighted average of synchronized dendro-somatic components of the input signals of a neural population. They find that BOLD correlates with LFP more than

MUA. They also record single-unit activity and find it to behave similarly to MUA. After linear systems modeling, Logothetis *et al.* describe the BOLD response to be a low pass filtered version of total neural activity.

The system of neurons, their oxygen consumption rate and the resulting smooth muscle response in the feeding blood vessels may be modeled mathematically with a system of differential equations as outlined in Section 4.2.1.

4.2.1 The balloon and other models

To a first approximation the hemodynamic response to a stimulus has been shown to be linear [64], which means that models of the form given by Equation (4.34) may be used. Nonlinear aspects of the hemodynamic response may be modeled using the Volterra kernel approach (see Section 4.1.2). The next step beyond a Volterra series approach is to model the physiology of the connection between neural activity and the hemodynamic response. These models usually begin with a model that relates the BOLD signal $y = \Delta S/S$ (see Equation (4.138)) to the normalized total deoxyhemoglobin $q = Q/Q_0$ and blood volume $v = V/V_0$, where Q_0 and V_0 are the rest quantities, via

$$y = V_0 \left[k_1(1 - q) + k_2(1 - q/v) + k_3(1 - v) \right]. \tag{4.141}$$

This model [76] assumes that the signal, S, originates from the post capillary venous vessels in the static averaging diffusion regime from both intravascular sources, S_i, and extravascular sources, S_e, weighted by blood volume fraction V as

$$S = (1 - V)S_e + VS_i. \tag{4.142}$$

Linearizing Equation (4.142) about V_0 gives

$$\Delta S = (1 - V_0)\Delta S_e - \Delta VS_e + V_0\Delta S_i + \Delta VS_i \tag{4.143}$$

and dividing Equation (4.143) by Equation (4.142) results in

$$\frac{\Delta S}{S} = \frac{V_0}{1 - V_0 + \beta V_0} \left[\frac{1 - V_0}{V_0} \frac{\Delta S_e}{S_e} + \frac{\Delta S_i}{S_e} + (1 - v)(1 - \beta) \right], \tag{4.144}$$

where $\beta = S_i/S_e$ is the intrinsic signal ratio at rest. The extravascular signal may be described by the following model based on numerical simulations by Ogawa *et al.* [349]:

$$\frac{\Delta S}{S} = -\Delta R_2^* T_E = aV_0(1 - q), \tag{4.145}$$

where $R_2^* = 1/T_2^*$ and $a = 4.3 \, \Delta\chi \, \omega_0 \, E_0 \, T_E$. At 1.5 T, with a susceptibility difference between the intra and extravascular space of $\Delta\chi = 1 \times 10^{-6}$, $\Delta\chi\omega_0 = 40.3$ s^{-1}. A resting oxygen extraction fraction of $E_0 = 0.4$ (see Equation (4.140)) and a $T_E = 40$ ms give $a = 2.8$. A model for the intravascular signal and the resting signal ratio β may be formulated from the experimental and simulation results of Boxerman et al. [53]. The model for the intravascular signal is

$$\frac{\Delta S}{S} = -\Delta A = -2\Delta E = 2\left(1 - \frac{q}{v}\right), \qquad (4.146)$$

where $A = 0.4 + 2(0.4 - E)$ for an oxygen extraction fraction in the range $0.15 < E < 0.55$ and $A_0 = \beta = 0.4$ assuming a magnetic field of 1.5 T, $T_E = 40$ ms and a blood vessel radius of 25 μm. Substituting Equations (4.145) and (4.146) into Equation (4.144) yields Equation (4.141) with $k_1 = a = 7E_0 = 2.8$, $k_2 = 2$ and $k_3 = 2E_0 - 0.2 = 0.6$ at 1.5 T with $T_E = 40$ ms and a vessel radius of 25 μm. The resting blood volume fraction per voxel is 0.01–0.04.

The balloon model postulates a coupled relationship between deoxyhemoglobin mass, Q, and venous volume V which can change according to pressures in a balloon like venous vessel that receives the output from a capillary bed. The coupled equations are

$$\frac{dQ}{qt} = F_{in}(t) E(F_{in}) C_a - F_{out}(V)\frac{Q(t)}{V(t)}, \qquad (4.147)$$

$$\frac{dV}{dt} = F_{in}(t) - F_{out}(V), \qquad (4.148)$$

where F_{in} is the CBF (ml/s) into the receiving venous vessel and F_{out} is the CBF out of the receiving venous vessel, C_a is the arterial oxygen concentration (in fully oxygenated blood) and E is the oxygen extraction fraction from the capillary bed before the blood enters the receiving venous vessel. In terms of normalized (dimensionless) variables $q = Q/Q_0$, $v = V/V_0$, $f_{in} = F_{in}/F_0$ and $f_{out} = F_{out}/F_0$, where the 0 subscripted variables are the resting values, Equations (4.147) and (4.148) become

$$\frac{dq}{qt} = \frac{1}{\tau_0}\left[f_{in}(t)\frac{E(f_{in})}{E_0} - f_{out}(V)\frac{q(t)}{v(t)}\right], \qquad (4.149)$$

$$\frac{dv}{dt} = \frac{1}{\tau_0}f_{in}(t) - f_{out}(v), \qquad (4.150)$$

where $\tau_0 = V_0/F_0$ is the mean transit time of blood through the receiving venous vessel at rest. The oxygen extraction fraction may be modeled as [75]

$$E = 1 - (1 - E_0)^{1/f_{in}} \qquad (4.151)$$

and the balloon like output may be described by a windkessel (elastic bag) model [296] as

$$f_{out} = v^{1/\alpha}. \qquad (4.152)$$

Mildner *et al.* [319] set $f_{in}E/E_0 = CMRO_2/CMRO_{2,0}$ (see Equation (4.140)) in Equation (4.149) in order to compare model predictions with their observations. In later work, Obata *et al.* [346] replace $E(f_{in})$ in Equation (4.149) (as represented by Equation (4.151)) with an independent function $E(t)$ to better represent the variation of BOLD and flow dynamics observed across brain regions.

Based on his experimental work, Glover [184] suggests that the windkessel model be modified to

$$f_{out} = 1 + \lambda_1(v - 1) + \lambda_2(v - 1)^\alpha, \tag{4.153}$$

where $\lambda_1 = 0.2$, $\lambda_2 = 4.0$ and $\alpha = 0.47$. Other experimental investigations have evaluated both the structure of the balloon model and the parameters in it. These experimental investigations combine fMRI with near-infrared spectroscopy (NIRS) for direct measurement of oxygenation [423], and with inversion recovery MRI [137] and exogenous contrast agent monocrystalline iron oxide nanocolloid (MION) [461] both to quantify CBF.

To connect the balloon model of Equations (4.149) and (4.150) and the BOLD model of Equation (4.141) to synaptic activity, Friston *et al.* [168] propose two additional equations to model the control of f_{in} by synaptic activity u as

$$\frac{df_{in}}{dt} = s(t), \tag{4.154}$$

$$\frac{ds}{dt} = \epsilon u(t) - s(t)/\tau_s - (f_{in}(t) - 1)/\tau_f, \tag{4.155}$$

where the new variable s can be thought to roughly represent a nitrous oxide (NO) astrocyte mediated signal from the synapse to the smooth muscles that control the vasodilation that controls f_{in}. The parameter ϵ represents the efficacy with which synapse activity causes an increase in the vasodilation signal, τ_s is the time constant for signal decay (or NO elimination) and τ_f is the time constant for auto-regulatory feedback from the blood flow. Tying together the NO signal model of Equations (4.154) and (4.155) with the balloon and BOLD models leads to a state space model that may be explicitly summarized as [148]

$$\frac{dx_i}{dt} = f_i(\vec{x}, u), \quad 1 \le i \le 4, \tag{4.156}$$

$$y = \lambda(\vec{x}), \tag{4.157}$$

where $\vec{x} = [x_1 \ x_2 \ x_3 \ x_4]^T$ is the "state space" vector (the unobserved variables of the model) and y is the BOLD response to the neural activity u in a single voxel.

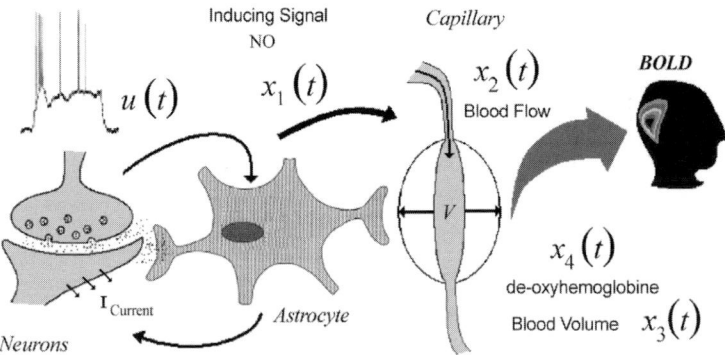

Fig. 4.5. Schematic of the meaning of the state space variables $x_1, \ldots, x_4$ of Equation (4.156). The BOLD signal is determined from the x_i through Equation (4.157). This diagram was taken from [374] and is used with permission.

The four functions f_i and λ are given explicitly by

$$f_1(\vec{x}(t), u(t)) = \epsilon u(t) - \kappa_s x_1(t) - \kappa_f [x_2(t) - 1], \tag{4.158}$$

$$f_2(\vec{x}(t), u(t)) = x_1(t), \tag{4.159}$$

$$f_3(\vec{x}(t), u(t)) = \frac{1}{\tau}[x_2(t) - x_3(t)^{1/\alpha}], \tag{4.160}$$

$$f_4(\vec{x}(t), u(t)) = \frac{1}{\tau}\left(x_2(t) \frac{[1 - (1 - E_0)^{1/x_2(t)}]}{E_0} - x_3(t)^{1/\alpha} \frac{x_4(t)}{x_3(t)} \right), \tag{4.161}$$

$$\lambda(\vec{x}(t)) = V_0 \left(7E_0[1 - x_4(t)] + 2\left[1 - \frac{x_4(t)}{x_3(t)}\right] + [2E_0 - 0.2][1 - x_3(t)] \right), \tag{4.162}$$

where Equations (4.158) and (4.159) correspond to Equations (4.154) and (4.155), Equations (4.160) and (4.161) represent the balloon model of Equations (4.149) and (4.150), and Equation (4.162) represents Equation (4.141). The state space variables explicitly are:

$x_1 = s = $ flow inducing signal caused by neural synaptic activity u;   (4.163)

$x_2 = f_{in} = $ normalized blood inflow into the voxel (CBF);   (4.164)

$x_3 = v = $ normalized blood volume in the voxel (CBV);   (4.165)

$x_4 = q = $ normalized deoxyhemoglobin mass.   (4.166)

A schematic of the meaning of these state space variables is given in Fig. 4.5 and some simulated versions of the state space functions are given in Fig. 4.6. Bayesian methods (see Section 4.4) may be used not only to solve for the unobserved

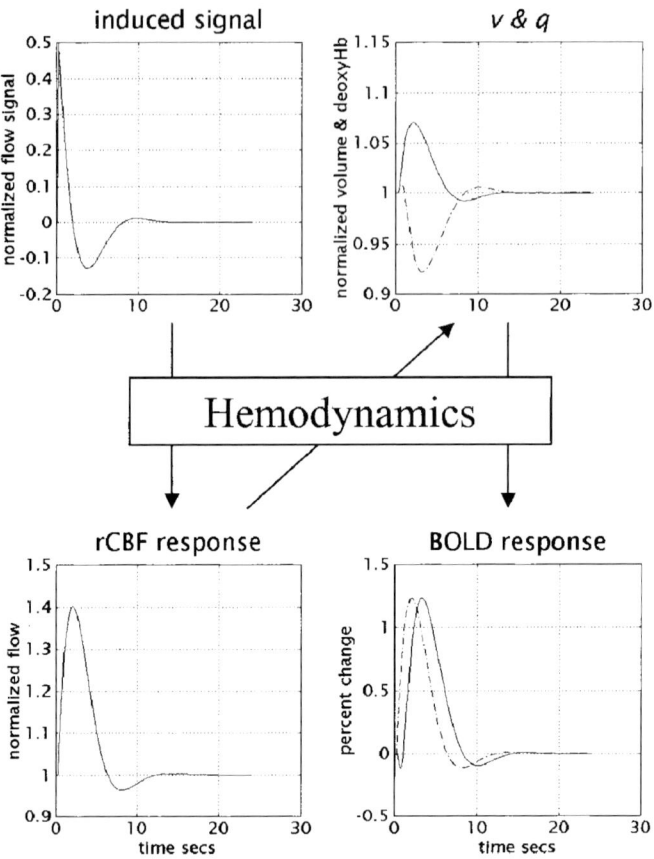

Fig. 4.6. Example time courses for the functions $x_1,\ldots,x_4$ and y of Equations (4.156) and (4.157) from a simulation study. Here induced signal = x_1, rCBF = x_2, $v = x_3$, $q = x_4$ and BOLD response = y. The dashed line in the BOLD response panel is x_2 and shows how the measured BOLD signal lags behind the blood flow response, which in turn lags behind the neural response modeled in this case as a short stimulus at $t = 0$. Note how the basic features of the BOLD response shown in Fig. 4.4 are reproduced by the modeled BOLD response here. This diagram was taken from [148] and is used with permission.

state variables but also for the parameters of the model. In particular, a map of ϵ may be considered to be an activation map. Complete state space models may also be used to generate model Volterra kernels which have then been favorably compared to measured Volterra kernels (see Section 4.1.4) [168]. Alternatively, the state space model of Equations (4.156)–(4.162) may be placed in a Bayesian framework (see Section 5.6.5) and posterior estimates of the state variables and their variances may be obtained pixel by pixel from an fMRI time-series [148, 374]. The extent and magnitude of the nonlinearities of the BOLD response for various experimental design parameters including prior stimulus, epoch length (for blocked designs), SOA for a fixed number of events and stimulus amplitude were investigated by Mechelli *et al.* [308] using simulations based on Equations (4.156) and (4.157).

Zheng *et al.* [474] (see also [213]) extend the state space model of Equations (4.156)–(4.162) to include a capillary bed model for oxygen transport to tissue (OTT) that replaces the model for the oxygen extraction fraction as given by Equation (4.151). To do this, $E(f_{in})$ in Equation (4.149) is replaced with a new state space variable $E(t)$ along with two more variables, $\overline{C}_B$, the spatial mean total blood O_2 concentration in the capillary bed that feeds the receiving venous vessel, and g, the ratio of the spatial average oxygen concentration in the tissue (C_P) to the plasma oxygen concentration in the arterial end of the capillary (C_P^a, so that $g = C_P/C_P^a$). The three new equations that describe E, $\overline{C}_B$ and g and their relationship to f_{in} and u are

$$\frac{dE}{dt} = \frac{f_{in}(t)}{\varphi}\left[-E(t) + (1 - g(t))\left(1 - \left(1 - \frac{E_0}{1 - g_0}\right)^{1/f_{in}(t)}\right)\right], \qquad (4.167)$$

$$\frac{d\overline{C}_B}{dt} = \frac{f_{in}(t)}{\varphi}\left[-\overline{C}_B(t) - \frac{C_B^a E(t)}{\ln(1 - E(t)/(1 - g(t)))} + C_B^a g(t)\right], \qquad (4.168)$$

$$\frac{dg}{dt} = \frac{1}{J}\frac{V_{cap}E_0}{V_{vis}rT}\left[\left(\frac{\overline{C}_B(t) - g(t)C_B^a}{\overline{C}_{B,0} - g_0 C_B^a} - 1\right) - Ku(t)\right], \qquad (4.169)$$

where the quantities with subscript 0 denote resting quantities, as usual, and φ is a constant related to capillary transit time, C_B^a is blood O_2 concentration at the arterial end of the capillary bed (taken to be 1 by Zheng *et al.*), V_{tis}, V_{cap} are the volume of blood in tissue and capillary respectively, $r = C_P/C_B$ is the ratio of plasma O_2 concentration, C_P, to total blood O_2 concentration C_B (r is assumed constant), T is the mean capillary transit time, J is a scaling constant, and K is a constant such that the metabolic demand M satisfies $M = Ku$.

Aubert and Costalat [22] formulate a detailed state space physiological model of the BOLD response using v_{stim} (in place of u) and F_{in} as input variables, with the BOLD response y as given by Equation (4.141) being the output. In their model v_{stim} represents stimulus induced sodium inflow into the cells (which, in principle, is a function of u) and there are a total of 15 variables in their model. Those variables are concentrations of: (1) intracellular sodium (Na_i^+), (2) intracellular glucose (GLC_i), (3) glyceraldehyde-3-phosphate (GAP), (4) phosphoenolpyruvate (PEP), (5) pyruvate (PYR), (6) intracellular lactate (LAC_i), (7) reduced form of nicotinamide adenine dinucleotide (NADH), (8) adenosine triphosphate (ATP), (9) phosphocreatine (PCr), (10) intracellular oxygen (O_{2i}), (11) capillary oxygen (O_{2c}), (12) capillary glucose (GLC_c), (13) capillary lactate (LAC_c), (14) venous volume ($V_v = V$ of Equation (4.148)) and (15) deoxyhemoglobin (dHb = Q of Equation (4.147)).

4.2.2 Neural activity from hemodynamic deconvolution

The state space models reviewed in Section 4.2.1 describe the HRF y in terms of neural activity u. A first approximation to u is to assume that it equals the stimulus function. Beyond that it is desired to deconvolve the cause of the HRF from the HRF to arrive at an estimated neural response u that is more useful for state space models and connectivity mapping (see Section 5.6). Although the hemodynamic response is known to be nonlinear, if relatively short presentations (SD, see Section 3.1) are used and the ISI (see Section 3.1) is long enough the response may be well approximated as linear[†]. With these assumptions, Glover [184] proposes deconvolving u from

$$y(t) = u * h(t) + \epsilon(t) \tag{4.170}$$

with an estimate u' of u obtained from a measurement of the IRF h using short duration SD in another experiment with the same subject and the following Wiener filter:

$$u'(t) = \mathcal{F}^{-1}\left[\frac{H^*(\omega)Y(\omega)}{|H(\omega)|^2 + E_0^2}\right], \tag{4.171}$$

where H and Y are the Fourier transforms of h and y, respectively and E_0 is an estimate of the assumed white noise spectrum amplitude of ϵ. A model for h for use in Equation (4.171) is obtained by fitting the function

$$h(t) = c_1 t^{n_1} e^{-t/t_1} - a_2 c_2 t^{n_2} e^{-t/t_2}, \tag{4.172}$$

with $c_i = \max(t^{n_i} e^{-t/t_i})$, to the data collected for IRF measurement. This method requires measurement of E_0 and is sensitive to the value assumed for the noise spectral density.

Gitelman *et al.* [183] generalize the approach to deconvolution in the following way. The discrete convolution obtained from evaluating Equation (4.170) at the sampled time points may be written in matrix form as

$$\vec{y} = [H]\vec{x} + \vec{\epsilon}, \tag{4.173}$$

where $[H]$ is a Toeplitz matrix representation of translated values of h. Let the columns of $[X]$ be a set of basis functions evaluated at the sampled time points;

[†] Janz *et al.* [231] show that nonlinearities from neural adaptation may also be responsible for nonlinearities in the BOLD response. That is, the neural response to the stimulus may become nonlinear as the neural system adapts to the stimulus.

Gitelman *et al.* use a Fourier basis set (cosines). Then Equation (4.173) becomes approximated by

$$\vec{y}_A = [H][X]\vec{\beta} + \vec{\epsilon}. \qquad (4.174)$$

By the Gauss–Markov theorem, the maximum likelihood estimator of $\vec{\beta}$ is

$$\hat{\vec{\beta}}_{ML} = ([X]^T[H]^T[\Sigma]^{-1}[H][X])^{-1}[X]^T[H]^T[\Sigma]^{-1}\vec{y}_A \qquad (4.175)$$

when $\vec{\epsilon} \sim N(0, \sigma^2[\Sigma])$. The high frequency components of $\hat{\vec{\beta}}_{ML}$ will generally be ruined by noise so the following maximum a-posteriori (MAP) estimate will be better:

$$\hat{\vec{\beta}}_{MAP} = ([X]^T[H]^T[\Sigma]^{-1}[H][X] + \sigma^2[C_\beta]^{-1})^{-1}[X]^T[H]^T[\Sigma]^{-1}\vec{y}_A, \qquad (4.176)$$

where $[C_\beta]^{-1}$ is the prior precision (inverse of the prior covariance) of $\vec{\beta}$. The diagonal elements of $[C_\beta]$ may be used to impose a smoothing constraint on the high frequency (or other) components of $\hat{\vec{\beta}}_{MAP}$. It can be shown that

$$\hat{\vec{\beta}}_{MAP} = \frac{H^*(\omega)Y(\omega)}{|H(\omega)|^2 + g_\epsilon(\omega)/g_\beta(\omega)}, \qquad (4.177)$$

where g_β is the prior spectral density and g_ϵ is the noise spectral density when Fourier basis vectors are used in $[X]$. It can be seen that Equation (4.177) is formally equivalent to Equation (4.171).

4.2.3 Hemodynamic and mental chronometry

For some studies it is desired to know the delay, or latency, of the BOLD response. Estimates of the time of the peak of the HRF may be used to infer timing of the underlying neural activity to $\sim$100 ms [66]. Also it has been shown that both the latency and width of the BOLD response correlate positively with the reaction time (RT) required by the subject to perform the given task [314]. Other features of the hemodynamic response that have been observed [78] are:

- The magnitude of the response is larger and the latency is longer for draining vessel voxels than for cortical voxels.
- Properties of the earlier part of the response, e.g. the early negative dip, provide better spatial localization than later parts of the response.
- The variability of the rising portion of the response is less than the variability of the descending portion.

One of the first approaches to measuring the latency of the BOLD response involves fitting a straight line to the ramp up of the BOLD response [313]. Working within ROIs identified as active using a conventional GLM for the IRF, Bellgowan *et al.* [39] model the measured HRF as a convolution between a Heavyside step function u, to represent the neural activity, with a gamma variate function to represent the hemodynamic response to neural activity. The width and delay of the Heavyside function is varied until a good fit between model and data is found, with the delay giving high resolution ~100 ms chronometry.

Another way of estimating the delay of the HRF is to fit a gamma density function of the form

$$f(t + l) = c \frac{b^a}{\Gamma(a)} (t + l)^{a-1} \exp(-b(x + l)) \qquad (4.178)$$

to the measured, mean subtracted, response, where l gives lag, the delay (the time to peak) is given by $l + a/b$ and c represents the response magnitude (area under the f). Hall *et al.* [204] uses such measures of lag, delay and magnitude plus a goodness-of-fit parameter in a logistic regression with a receiver operating curve (ROC) analysis to show that such parameters may be used to distinguish between voxels containing gray matter and those containing primarily veins with a discriminating power (area under the ROC curve) of 0.72.

Saad *et al.* [381] use correlation analysis to both detect activation and quantify latency. They compute the cross-correlation function between the measured voxel time-series $y(t)$ and scaled and translated versions of a reference function $x(t)$ given by $\alpha x(t - \delta)$. (Their reference function is a simple sine wave having a frequency equal to the frequency of their block design experiment.) The resulting cross-correlation function r_{xy} is related to the autocorrelation function r_{xx} of x by

$$r_{xy}(\tau) = \alpha r_{xx}(\tau - \delta), \qquad (4.179)$$

which will be maximum when $\tau = \delta$. The corresponding correlation coefficient $r(\delta)$, given by

$$r(\delta) = \frac{r_{xy}(\delta)}{\sqrt{r_{xx}(0) r_{yy}(0)}}, \qquad (4.180)$$

may be used as a thresholding parameter for detecting activation. The maximum of r_{xy} occurs where the Hilbert transform of r_{xy}, Hr_{xy}, equals zero. Finding the zero of Hr_{xy} is computationally more efficient[†] than finding the maximum of r_{xy}.

Calhoun *et al.* [78] use a weighted least squares approach to measure latency, which they define essentially as time to onset by using a fixed model for the HRF that may be shifted by δ. For each δ, the GLM

$$\vec{y} = [X_\delta] \vec{\beta} + \vec{\epsilon}, \qquad (4.181)$$

[†] Sadd *et al.* [381] report that their software is available as a plug-in for the AFNI software.

where the column of interest in $[X_\delta]$ is the fixed HRF shifted by δ, may be solved using

$$\vec{\hat{\beta}}_\delta = ([X_\delta]^T [W][X_\delta])^{-1} [X_\delta]^T [W]\vec{y}, \qquad (4.182)$$

where $[W]$ is a weighting matrix that is used to more heavily weight the earlier response time points to account for the reduced variability that has been observed in the earlier parts of the HRF. Then, for each δ, an adjusted correlation coefficient, ρ_δ, is computed as

$$\rho_\delta = \sqrt{\frac{(N'-1)R_\delta^2 - K}{N' - K - 1}}, \qquad (4.183)$$

where K is the number of explanatory variables, N' is the effective degrees of freedom (see Section 4.1.3) and

$$R_\delta^2 = \frac{\vec{\hat{\beta}}[X_\delta]^T [W]^T \vec{y} - \vec{y}^T \text{diag}([W])^2 / \text{diag}([W])^T [I]}{\vec{y}^T [W]\vec{y} - \vec{y}^T \text{diag}([W])^2 / \text{diag}([W])^T [I]}. \qquad (4.184)$$

The δ that maximizes ρ_δ defines the latency. Calhoun *et al.* compute ρ_δ in the Fourier domain to increase computational efficiency and resolution on values of δ considered.

Friston *et al.* [161] use two basis functions to represent the IRF as given by Equation (4.35), a sum of two gamma variate functions as given by Equation (4.195) and its temporal derivative. Once β_b of Equation (4.35) are determined, then some aspect of the inferred IRF h, such as the time of the maximum response, may be used as a measure of latency l. Friston *et al.* show that the standard error of such a latency estimate is given by

$$SE(l) = SE(h(\tau)) \left/ \left(\frac{dh}{dt}(\tau) \right), \right. \qquad (4.185)$$

where τ is the time associated with the latency parameter ($l = \tau$ for the example of time of peak relative to stimulus onset). The logic of using the sum of a model IRF and its derivative to model latency goes as follows. Let $h_0(t)$ be a model IRF given by, for example, Equation (4.195), then define a shifted version of h_0 by

$$h(t; \delta) = h_0(t - \delta) \qquad (4.186)$$

so that

$$h(t; \delta) \approx h_0(t) + h_1(t)\delta \qquad (4.187)$$

by Taylor's theorem, where

$$h_1(t) = \left. \frac{\partial h(t; \delta)}{d\delta} \right|_{\delta=0} = -\frac{dh_0}{dt}(t). \qquad (4.188)$$

Convolving with the stimulus paradigm u gives two functions $x_0 = u * h_0$ and $x_1 = u * h_1$ and a GLM that may be expressed as

$$y_i = x_0(t_i)\beta_1 + x_1(t_i)\beta_2 + \epsilon_i \qquad (4.189)$$

$$= u * (\beta_1 h_0(t_i) + \beta_2 h_1(t_i)) + \epsilon_i \qquad (4.190)$$

$$\approx u * h(t_i; \delta)\alpha + \epsilon_i, \qquad (4.191)$$

where α is a scaling factor. Comparison with Equation (4.187) yields $\delta = \beta_2/\beta_1$ [216]. Liao *et al.* [277] extend this approach by picking different two basis functions b_0 and b_1 so that

$$h(t; \delta) \approx b_0(t)w_0(\delta) + b_1(t)w_1(\delta) \qquad (4.192)$$

is a better approximation than Equation (4.187). Convolving with the stimulus paradigm u gives two functions $x_0 = u * b_0$ and $x_1 = u * b_1$ and a GLM that may be expressed as

$$y_i = x_0(t_i)w_0(\delta)\gamma + x_1(t_i)w_1(\delta)\gamma + \epsilon_i, \qquad (4.193)$$

so that the parameter vector is $\vec{\beta} = [w_0(\delta)\gamma \; w_1(\delta)\gamma]^T$. Once $\vec{\beta}$ is estimated, the parameter of interest, δ, and an associated t statistic can be computed.

Hernandez *et al.* [217] explore, using simulation and experiment, the effect of sequence and stimulus timing errors on the determination of BOLD delay. They let $B(t)$ be the BOLD actual response (HRF), let $y(t) = B(t) + \epsilon(t)$ be the signal model with added noise and let $x(t) = B(t - T) \equiv B_T(t)$ represent an analysis function in error by an offset of T and derive the correlation between x and y, r_{xy} to be

$$r_{xy} = \frac{\langle B, B_T \rangle - \bar{B}^2}{\sigma_B \sqrt{\sigma_\epsilon^2 + \sigma_B^2}} \qquad (4.194)$$

assuming $\bar{B} = \bar{B_T}$. Note that $\langle B, B_T \rangle$ equals the autocorrelation function of B evaluated at $-T$ so that r_{xy} is dominated by the behavior of that autocorrelation function, and hence on the form of B. For their simulations an IRF of the form

$$r(t) = \frac{\ell_1^{\tau_1} t^{(\tau_1-1)} e^{(-\ell_1 t)}}{\Gamma(t_1)} - \frac{\ell_2^{\tau_2} t^{(\tau_2-1)} e^{(-\ell_2 t)}}{\Gamma(t_2)}, \qquad (4.195)$$

where Γ is given by Equation (3.6) and $\tau_1 = 6$, $\tau_2 = 16$, $\ell_1 = 1$, $\ell_2 = 1$, was used (time units in seconds). The model of Equation (4.195) is an improvement over that of Equation (3.4) in that it is capable of representing the undershoot, after response, of the BOLD function. By looking at the corresponding t statistic

$$t = r_{xy} \sqrt{\frac{v}{1 - r_{xy}^2}}, \qquad (4.196)$$

where ν is the degrees of freedom, Hernandez *et al.* show that, as a function of T, the statistic t has a relatively sharp peak at typical fMRI SNRs so that a series of analyses that vary the temporal offset of the model HRF would be able to arrive at the correct offset by maximizing t with respect to T.

4.3 Other parametric methods

The GLM is widely used for the analysis of fMRI time-series, and is easily the most popular method, but other approaches have been tried and are being used on a regular basis by some research groups. Here we review other approaches that may be characterized in terms of estimating parameters associated with the BOLD response.

One interesting early approach to activation map computation for a blocked design uses likelihood ratios formed over clusters of pixels to increase SNR [378]. In that approach, the SNR at pixel p is defined in the usual way as $\mathrm{SNR}_p = s/\sigma$, where s is the signal and σ is the standard deviation of the noise. Over a region of N pixels, the SNR increases to $\mathrm{SNR}_r = \sqrt{N}\,\mathrm{SNR}_p$. Signal in an active region of voxels is assumed to have a mean of μ_1 while a nonactive region has a mean of μ_0 and the probability of observing a value y in each case is assumed to be

$$p_i(y)\frac{1}{\sigma\sqrt{2\pi}}e^{-(y-\mu_i)^2/2\sigma^2}, \tag{4.197}$$

which is a good assumption given the long T_R of their investigation (small noise autocorrelation). For a time-series of region values $\vec{y}$ of length n containing r rest time points, the likelihood ratio, L, is shown to be

$$L(\vec{y}) = \prod_{k=r+1}^{n} \frac{p_1(y_k)}{p_0(y_k)} \tag{4.198}$$

and a region is considered active if $L(\vec{y}) > \gamma$ for $\ln \gamma = (n-r)\ln \alpha$, where α represents a decision level for rejecting the null hypothesis such that $[p_1(y_k)/p_0(y_k)] > \alpha$ for each k. Regions are initially defined by computing likelihood ratios for each voxel's time-series. Then adjacent activated voxels define the regions for the next likelihood ratio calculation. Regions are then trimmed on the basis that the individual voxels, p, have to have an SNR that passes the test

$$\mathrm{SNR}_p > \frac{1}{\sqrt{N-1}}\mathrm{SNR}_r. \tag{4.199}$$

A similar likelihood ratio test, at the single-voxel level, has been used on complex image data[†] to provide more sensitive detection [335].

[†] MRI data are intrinsically complex and typically only the magnitude image is provided after Fourier transforming the raw data. Here it is assumed that the complex data are available for analysis.

An ANOVA-based GLM approach to blocked designs, BOLDfold, has been proposed [97, 389]. Let there be N epochs with K time points (scans) in each epoch to give a total of $n = NK$ data volumes in the time-series. Denote the scan times as t_{ij}, where i indexes the epoch and j indexes the time within the epoch. Group k of the ANOVA is defined by the time points $\{t_{ik} | 1 \le i \le N\}$ so that there are K groups. The average of each group, taken in order, then represents the average BOLD response over the epochs and may be tested for flatness by comparing the standard

$$F = \frac{\mathrm{MS}_g}{\mathrm{MS}_e} \tag{4.200}$$

to F_{ν_1, ν_2}, where MS_g is the mean square difference between groups and MS_e is the mean square error, and $\nu_1 = K - 1$ and $\nu_2 = K(N - 1)$ are the treatment and error degrees of freedom, respectively. It may be shown that

$$F = \frac{\nu_2}{\nu_1} \frac{r^2}{(1 - r^2)}, \tag{4.201}$$

where r is the correlation between the measured time-series and the average response repeated N times given explicitly by

$$r^2 = \frac{\left[\sum_{j=1}^{N} \sum_{i=1}^{K} (\bar{y}_{i\cdot} - \bar{y}_{\cdot\cdot})(y_{ij} - \bar{y}_{\cdot\cdot}) \right]^2}{\left[N \sum_{i=1}^{K} (\bar{y}_{i\cdot} - \bar{y}_{\cdot\cdot})^2 \right] \left[\sum_{j=1}^{n} \sum_{i=1}^{k} (y_{ij} - \bar{y}_{\cdot\cdot})^2 \right]}, \tag{4.202}$$

where $\bar{y}_{\cdot\cdot}$ represents the grand mean of the time-series values and $\bar{y}_{i\cdot}$ represents the mean of group i [389]. A similar approach is proposed by Lu *et al.* [289]. If we define $\vec{Y}_i = [y_{i1} \cdots y_{iK}]^T$ and the correlation between $\vec{Y}_i$ and $\vec{Y}_j$ as c_{ij}, then a map of T parameters may be thresholded to give an activation map where

$$T = \frac{(2/N(N-1)) \sum_{i=1}^{N} \sum_{j>i} c_{ij}}{\sqrt{(2/N(N-1)) \sum_{i=1}^{N} \sum_{j>i} (c_{ij} - (2/N(N-1)) \sum_{i=1}^{N} \sum_{j>i} c_{ij})^2}}. \tag{4.203}$$

These ANOVA type approaches have the advantage of providing a model free estimate of the HRF (as opposed to the IRF) in blocked designs. A similar model free approach, not requiring a blocked design, but requiring multiple identical imaging runs, may be obtained by computing the correlation coefficients of time courses in corresponding voxels between the runs [276].

Purdon *et al.* [366] and Solo *et al.* [406] construct a physiologically motivated model, where the signal is assumed proportional to blood volume times deoxyhemoglobin concentration VQ, in which spatiotemporal regularization (smoothing[†]) is

---

[†] A local polynomial technique is used in the regularization procedure in lieu of a computationally intractable global Tikhonov regularization procedure.

used to estimate the model parameters. Their model for the measured data $x(t,\vec{p})$ at time point t and pixel[†] $\vec{p}$ may be expressed as

$$x(t,\vec{p}) = m(\vec{p}) + b(\vec{p})t + s(t,\vec{p}) + v(t,\vec{p}), \tag{4.204}$$

where $m(\vec{p}) + b(\vec{p})t$ is a drift correction term, $s(t,\vec{p})$ is the signal model and $v(t,\vec{p})$ is the noise model. The signal is modeled as $s(t,\vec{p}) = V(t,\vec{p})Q(t,\vec{p})$ where

$$Q(t,\vec{p}) = k_1(\vec{p}) + k_2(\vec{p})(g_a * u)(t), \tag{4.205}$$

$$g_a(t) = (1 - e^{-1/d_a})^2 (t + 1) e^{-t/d_a} \tag{4.206}$$

and

$$V(t,\vec{p}) = k_3(\vec{p}) + k_4(\vec{p})(g_b * u)(t), \tag{4.207}$$

$$g_b(t) = (1 - e^{-1/d_b}) e^{-t/d_b}, \tag{4.208}$$

where g_a is a gamma variate function, $k_1(\vec{p}), \ldots, k_4(\vec{p})$ are constants, $d_a = 1.5$ s, $d_b = 12$ s and u is the stimulus function. So

$$s(t,\vec{p}) = f_a(\vec{p})(g_a * u)(t - \delta(\vec{p})) + f_b(\vec{p})(g_b * u)(t - \delta(\vec{p}))$$
$$+ f_c(\vec{p})(g_a * u)(t - \delta)(g_b * u)(t - \delta(\vec{p})), \tag{4.209}$$

where $\vec{\beta}(\vec{p}) = [m(\vec{p}) b(\vec{p}) f_a(\vec{p}) f_b(\vec{p}) f_c(\vec{p}) \delta(\vec{p})]^T$ are the signal parameters to be estimated. The noise model is given by

$$v(t,\vec{p}) = w(t,\vec{p}) + \eta(t,\vec{p}), \tag{4.210}$$

$$w(t,\vec{p}) = \rho(\vec{p})w(t-1,\vec{p}) + \epsilon(t,\vec{p}), \tag{4.211}$$

where $\eta(t,\vec{p})$ is zero-mean Gaussian white noise with variance $\sigma_\eta^2(\vec{p})$ representing scanner noise, $\epsilon(t,\vec{p})$ is zero-mean Gaussian white noise with variance $\sigma_\epsilon^2(\vec{p})$ and $\rho(\vec{p})$ is a correlation coefficient; $w(t,\vec{p})$ is an AR(1) model and the model is equivalent to an ARMA(1,1) model [52]. So the noise parameters to be estimated are $\vec{\alpha}(\vec{p}) = [\sigma_\eta^2(\vec{p}) \sigma_\epsilon^2(\vec{p}) \rho(\vec{p})]^T$. Altogether, the parameters to be estimated are given by $\vec{\theta}(\vec{p}) = [\vec{\beta}^T \vec{\alpha}^T]^T$. The parameters are estimated by minimizing

$$J(\vec{\theta}) = J([\vec{\theta}(1,1)^T \ldots \vec{\theta}(M_1, M_2)^T]^T) = \sum_{\vec{p}} J_{\vec{p}}(\vec{\theta}(\vec{p})), \tag{4.212}$$

where the criterion at pixel $\vec{p}$ is a spatially locally weighted log-likelihood given by

$$J_{\vec{p}}(\vec{\theta}(\vec{p})) = \sum_{\vec{q}} K^h(\vec{p} - \vec{q}) L(\vec{q}; \vec{\theta}(\vec{p})), \tag{4.213}$$

[†] The spatial regularization used is 2D but could, in principle, be extended to 3D.

where $L(\vec{q}; \vec{\theta}(\vec{p}))$ is a Gaussian log-likelihood based only on the time-series at pixel $\vec{q}$ and $K^h(\vec{p}-\vec{q})$ is a spatial smoothing kernel. The Gaussian log-likelihood at pixel $\vec{q}$ is given in terms of the temporal discrete Fourier transforms (DFT) $F[x(t,\vec{q})] = \tilde{x}(k,\vec{q})$ and $F[\mu(t,\vec{p};\vec{\beta}(\vec{p}))] = F[m(\vec{p}) + b(\vec{p})t + s(t,\vec{p})] = \tilde{\mu}(k,\vec{p};\vec{\beta}(\vec{p}))$

$$L(\vec{q}; \vec{\theta}(\vec{p})) = -\frac{1}{2}\sum_k \frac{|\tilde{x}(k,\vec{q}) - \tilde{\mu}(k,\vec{p};\vec{\beta}(\vec{p}))|^2}{N^2 S(k;\vec{\alpha}(\vec{p}))} - \frac{1}{2}\sum_k \ln S(k;\vec{\alpha}(\vec{p})), \quad (4.214)$$

where N is the number of time points and $S(k;\vec{\alpha}(\vec{p}))$ represents the noise spectrum. The Epanechnikov smoothing kernel is given by

$$K^h(\vec{q}) = K\left(\frac{q_1}{M_1 h}\right) K\left(\frac{q_2}{M_2 h}\right) \frac{1}{h^2 M_1 M_2}\left(1 - \frac{1}{4M_1^2}\right)\left(1 - \frac{1}{4M_2^2}\right), \quad (4.215)$$

where M_1 and M_2 are the image dimensions and

$$K(x) = \begin{cases} 3(1 - x^2)/4 & \text{for } |x| \le 1 \\ 0 & \text{for } |x| > 1 \end{cases}. \quad (4.216)$$

Equation (4.213) is minimized by an iterative procedure (see [406]) to complete the process which is called locally regularized spatiotemporal (LRST) modeling.

Purdon *et al.* [366] and Solo *et al.* [406] also compare their model to the standard GLM using the nearly unbiased risk estimator (NURE) technique. This technique is based on an estimate of the Kullback–Leiber (KL) distance, $R(\theta)$, between the true probability of the data, $p(x)$, and the probability density given by the model, $p_\theta(x)$, defined by

$$R(\theta) = \int p(x) \ln\left(\frac{p(x)}{p_\theta(x)}\right) dx \quad (4.217)$$

plus a measure of the number of parameters in the model. As always, there are trade-offs between models but Purdon *et al.* and Solo *et al.* find that LRST outperforms the standard GLM in terms of modeling the noise better than temporal filtering (see Section 4.1.1) and fewer parameters are required to model nonlinearities than in a Volterra approach (see Section 4.1.4). However, the GLM approach requires considerably less computation.

Ledberg *et al.* [272] take a 4D approach to activation map computation. In their approach multiple data runs are required with the assigned task being constant within a run. Activations are determined by comparing runs in a GLM. Ledberg *et al.* recommend that a large number of short runs be used as opposed to the standard approach of using a small number of long runs. Let there be t time points in each run, v voxels in each volume and n runs in total and organize the data into an $n \times vt$ data matrix $[Y]$ as follows:

$$[Y] = \begin{bmatrix} \vec{v}_{11}^T & \cdots & \vec{v}_{1t}^T \\ \vec{v}_{21}^T & \cdots & \vec{v}_{2t}^T \\ & \cdots & \\ \vec{v}_{n1}^T & \cdots & \vec{v}_{nt}^T \end{bmatrix}, \qquad (4.218)$$

where $\vec{v}_{ij}^T$ is the volume at time j in run i, so each row represents a run. The "4D" GLM is then

$$[Y] = [X][B] + [E], \qquad (4.219)$$

where $[X]$ is an $n \times p$ design matrix, $[B]$ is a $p \times vt$ matrix of model parameters and $[E]$ is an $n \times vt$ matrix of error terms. The rows of the design matrix code for the conditions represented by the runs (e.g. the runs could alternate between task and rest) and the columns represent volumes at a given time in every run. If the expected value of $[E]$ is $[0]$ and $\text{cov}([e]_j) = \sigma_j^2[I]$, where $[e]_j$ is the $n \times v$ matrix that represents the jth column of $[E]$ and $\vec{h}$ is a p-dimensional estimable contrast (i.e. $\vec{h}^T = \vec{a}^T[X]$ for some $\vec{a}$), then the BLUE estimate of $\vec{h}^T[B]$ is given by

$$\vec{h}^T[\hat{B}] = \vec{h}^T([X]^T[X])^+[X]^T[Y] \qquad (4.220)$$

with an associated 4D statistical t vector of dimension vt given by

$$\vec{t} = (\vec{h}^T[\hat{B}])(\text{diag}([R]^T[R])\vec{h}^T([X]^T[X])^+\vec{h}/\gamma)^{-1/2}, \qquad (4.221)$$

where $[R] = [Y] - [X][\hat{B}]$ is the matrix of residuals, $\gamma - n - \text{rank}([X])$ is the degrees of freedom of the model and diag is the matrix operation that sets nondiagonal entries to zero. Inferences can be made at the 4D voxel level or 4D cluster level by using permutation or Monte Carlo techniques to determine the null hypothesis distributions. The voxel level inferences yield vt size activation maps (movies) when the values of $\vec{t}$ are thresholded at some level.

Katanoda et al. [240] modify the GLM to base parameter estimation for a voxel on the time courses of neighboring voxels. They illustrate their approach with a simple regression model in which the design matrix has one column so that with the standard GLM approach the model at voxel u_0 is given by

$$y(t, u_0) = x(t)\beta_{u_0} + \epsilon(t, u_0). \qquad (4.222)$$

The standard model is extended to include neighboring voxels and expressed as

$$\begin{bmatrix} y(t, u_0) \\ \cdots \\ y(t, u_\ell) \end{bmatrix} = \begin{bmatrix} x(t) \\ \cdots \\ x(t) \end{bmatrix} \beta_{u_0} + \begin{bmatrix} \epsilon(t, u_0) \\ \cdots \\ \epsilon(t, u_\ell) \end{bmatrix}, \qquad (4.223)$$

where $u_1, \ldots, u_\ell$ represent neighboring (spatial) voxels. Katanoda *et al.* solve the model of Equation (4.223) in the temporal Fourier domain where the temporal auto-correlations (AR(1) model) and spatial correlations that are specified in a separable fashion are more easily modeled. The Fourier domain approach leads to a model that can be solved by an ordinary least squares method.

4.3.1 Nonlinear regression

The GLM is a linear regression approach but nonlinear regression may also be used to detect activations and quantify the HRF. Kruggel and von Cramon [258] propose a nonlinear regression approach by first dividing the fMRI volume time-series into a set of k ROIs, S, (defined on the basis of a preliminary GLM analysis) and time points $T = \{1, \ldots, \ell\}$, where ℓ is the number of time points, so that the reduced data set for model fitting is $\{y(s, t) | s \in S, t \in T\}$. The nonlinear model used is

$$y(s, t) = g(t, \vec{\beta}) + \epsilon(s, t) \tag{4.224}$$

for a total of $n = k\ell$ equations, where it is assumed that $\epsilon \sim N_n(0, [V])$ and $\vec{y} \sim N_n(\vec{g}(T, \vec{\beta}), [V])$ with N_n denoting the n-variate normal distribution. The model, per se, is defined by

$$g(t, \vec{\beta}) = \frac{\beta_0}{\beta_1} \exp \left(-\frac{(t - \beta_2)^2}{2\beta_1^2} \right) + \beta_3, \tag{4.225}$$

where β_0 is interpreted as gain (amplitude), β_1 as dispersion (width), β_2 as lag and β_3 as the baseline. Starting with $[V] = [I]$ the parameters $\vec{\beta}$ are computed by minimizing

$$\arg \min_{\vec{\beta}} \left[\vec{y} - \vec{g}(T, \vec{\beta})^T [V]^{-1} \vec{g}(T, \vec{\beta}) \right] \tag{4.226}$$

using the downhill simplex method. Then the residuals are used to compute $[V]$ for use in the next iteration of optimizing Equation (4.226). After five or so iterations the values for $\vec{\beta}$ are adopted, the residuals checked for normality and stationarity and confidence intervals computed.

4.4 Bayesian methods

The GLM and other methods described thus far are applied by considering the null hypothesis, H_0, of no activation and looking for evidence against H_0 in a stat-istic with a low p value, the probability of incorrectly rejecting H_0. This approach is known as the frequentist approach[†]. Recently, many investigators have been

---

[†] Many authors in the fMRI literature refer to the frequentist approach as the "classical" approach even though Bayes's ideas were developed in the 1700s and Fisher promoted the frequentist approach in the early

developing Bayesian methods as an alternative to frequentist methods. In this section we review Bayesian approaches to making inferences about activation. In Section 5.6.5 we review Bayesian approaches to making inferences about connectivity. The boundary between Bayesian activation inference and Bayesian connectivity inference is a fuzzy one with the boundary being filled by the application of multivariate and/or clustering methods in a Bayesian way.

To illustrate the ideas of the Bayesian approach, following Woolrich *et al.* [450] (see also [41]), let Θ represent the parameters (a set) of a given model $\mathcal{M}$ (a set of equations) and let Y represent the data (a vector for univariate voxel analysis or a data matrix for multivariate approaches). Then Bayesian methods are based on the Bayes rule (see Equation (2.17))

$$p(\Theta|Y, \mathcal{M}) = \frac{p(Y|\Theta, \mathcal{M}) \, p(\Theta|\mathcal{M})}{p(Y|\mathcal{M})}, \qquad (4.227)$$

where each term represents a probability distribution[†]. For any given data set, the term $p(Y|\mathcal{M})$ is constant so we may absorb it into the function $\ell(\Theta|Y, \mathcal{M}) = p(Y|\Theta, \mathcal{M})/p(Y|\mathcal{M})$ and write

$$p(\Theta|Y, \mathcal{M}) = \ell(\Theta|Y, \mathcal{M}) \, p(\Theta|\mathcal{M}), \qquad (4.228)$$

where $p(\Theta|\mathcal{M})$ is the *prior* distribution for Θ, $\ell(\Theta|Y, \mathcal{M})$ is the *likelihood function* and $p(\Theta|Y, \mathcal{M})$ is the *posterior* distribution. A computational complication in Equation (4.228) is that

$$p(Y|\mathcal{M}) = \int_{\Theta} p(Y|\Theta, \mathcal{M}) \, p(\Theta|\mathcal{M}) \, d\Theta \qquad (4.229)$$

needs to be computed and the integral is frequently not tractable analytically. Integrals describing the marginal posterior distributions for parameters of interest,

$$p(\Theta_I|Y, \mathcal{M}) = \int_{\Theta_{\neg I}} p(\Theta|Y, \mathcal{M}) \, d\Theta_{\neg I}, \qquad (4.230)$$

where Θ_I and $\Theta_{\neg I}$ are the parameters of interest and noninterest respectively, also need to be computed and also may not be tractable analytically. Without analytical solutions, the integrals need to be estimated numerically. A popular way of computing such integrals numerically is via a Markov chain Monte Carlo (MCMC) [174, 182] method, where samples of Θ are generated from the given distribution in such a way that a histogram of those samples matches the given probability distribution. The MCMC method requires many samples for an accurate estimate of the given probability distribution integral but one can get away with fewer samples

---

1900s. From the point of view of the history of fMRI analysis (which began in the early 1990s), however, the frequentist approach may be considered "classical".

[†] A probability distribution may be defined as a non-negative function in $L^2(\mathbb{R}^n)$ whose norm is 1.

by fitting a Student t distribution to the sample histogram in a method known as Bayesian inference with a distribution estimation using a T fit (BIDET).

Frequentist approaches find Θ that maximize the likelihood ℓ, frequently using least squares methods (e.g. Equation (4.79)) or using the iterative *expectation maximization* (EM) method described later in this section. That is, the parameters computed in frequentist methods are ML parameters. Bayesian methods find the MAP parameters, i.e. the parameters that maximize $p(\Theta|Y, \mathcal{M})$. When the prior distribution is flat (constant[†]), the ML solution and MAP solutions coincide. The advantage to the Bayesian approach is that a-priori information may be used in making inferences. This means that one need not always start "from scratch", essentially discounting previous knowledge and wisdom, when making inferences from a given data set. Knowledge from previous experiments could, in principle, be used as a-priori information in a Bayesian approach. Frequently, the a-priori information is in terms of the smoothness (regularization approach) of the response or in terms of the structure of the noise variance components (e.g. the autocorrelation structure of the fMRI time series would be modeled in the prior term). Other differences derive from the differing inference philosophies of the frequentist and Bayesian approaches. With the frequentist approach, one states the probability that an activation could be *false* (the p value) on the basis of an a-priori assumed (or sometimes estimated by permutation methods) null hypothesis probability distribution. With the Bayesian approach, one states the probability that an activation is *true*. This difference means that there is no multiple comparison problem with Bayesian methods; the posterior probabilities do not need to be adjusted in the same way that p values do (see Section 4.1.8). However, a minimum threshold probability must still be chosen to define the boundary between active and nonactive voxels.

4.4.1 Bayesian versions of the GLM

The first applications of Bayesian theory to fMRI activation map computation were made by Frank *et al.* [143] and by Kershaw *et al.* [243]. This work began with the GLM as given by

$$\vec{y} = [A]\vec{\theta} + [B]\vec{\psi} + \vec{\epsilon} = [X]\vec{\beta} + \vec{\epsilon}, \qquad (4.231)$$

where $[X] = [[A]|[B]]$ and $\vec{\beta} = [\vec{\theta}^T \, \vec{\psi}^T]^T$ gives a partition of the design matrix and parameter vector into effects of interest and no interest, respectively, with $\dim \vec{\theta} = m$ and $\dim \vec{\psi} = q$. Assuming iid normally distributed variation in $\vec{\epsilon}$,

---

[†] A flat distribution is an *improper* distribution because its integral is infinite. However, an improper distribution multiplied with a proper distribution gives a proper distribution.

Equation (4.231) leads to the likelihood function

$$\ell(\vec{y}|\vec{\beta},\sigma^2) = \frac{K}{(2\pi\sigma^2)^n} \exp\left[-\frac{(\vec{y}-[X]\vec{\beta})^T(\vec{y}-[X]\vec{\beta})}{2\sigma^2}\right], \tag{4.232}$$

where K is a normalization constant, $n = \dim \vec{y}$ and σ is a *hyperparameter* that quantifies the variance and which needs to be estimated along with the parameters in $\vec{\beta}$. Since nothing is known about the distribution of the parameters a-priori, a *noninformative* prior distribution may be used. Noninformative priors may be constructed using Jefferys's Rule which states that for a set of nonindependent parameters, $\vec{\Omega}$, the noninformative prior may be taken as

$$p(\vec{\Omega}) \propto \sqrt{\det[J]}, \tag{4.233}$$

where $[J]$ is the Fisher information matrix having entries

$$J_{ij} = E\left[\frac{-\partial^2(\ln\ell)}{\partial\Omega_i\,\partial\Omega_j}\right]. \tag{4.234}$$

Assuming that the variance hyperparameter is independent of the other parameters, $p(\vec{\beta},\sigma^2) = p(\vec{\beta})p(\sigma^2)$, applying Jefferys's Rule leads to the noninformative prior distribution $p(\vec{\beta},\sigma^2) = 1/\sigma^2$. In this case, the normalization integral may be computed analytically to give a posterior distribution of

$$p(\vec{\beta},\sigma^2|\vec{y}) = \frac{(rs^2)^{r/2}\sqrt{\det([X]^T[X])}\,(\sigma^2)^{-(p/2+1)}}{2^{p/2}\,\Gamma(1/2)^{m+q}\,\Gamma(r/2)} \exp\left[-\frac{rs^2+Q(\vec{\beta})}{2\sigma^2}\right], \tag{4.235}$$

where

$$r = n - m - q, \quad Q(\vec{\beta}) = (\vec{\beta} - \hat{\vec{\beta}})^T[X]^T[X](\vec{\beta} - \hat{\vec{\beta}})$$

and

$$\hat{\vec{\beta}} = ([X]^T[X])^{-1}[X]^T\vec{y}.$$

The marginal posterior probability density for the parameters of interest may also be analytically computed. From Equation (4.235) one may compute the MAP estimate of $(\vec{\beta},\sigma^2)$ or the expectation value. Both methods lead to the ML value $\hat{\vec{\beta}}$ but the MAP calculation produces $p\hat{\sigma}^2/(p+2)$ and the expectation value calculation produces $p\hat{\sigma}^2/(r-2)$ as estimates for σ^2, where $\hat{\sigma}^2$ is the ML value (the variance of $\vec{\epsilon}$) so that each method gives a different posterior probability.

Both Goutte *et al.* [195] and Marrelec *et al.* [298] use the finite impulse response (FIR) model of the IRF (given by Equation (4.37)) as the design matrix for the parameters of interest and a model of the error autocovariance $[\Sigma]$ in the prior via the matrix $[R] = [\Sigma]^{-1}$, modified to constrain the IRF to begin and end at zero, to result in MAP estimate of

$$\vec{\beta}_{MAP} = ([X]^T[X] + \sigma^2[R])^{-1}[X]^T\vec{y}, \tag{4.236}$$

where the hyperparameter σ^2 may be obtained as the value that maximizes the evidence $p(\vec{y}|[X], \sigma^2, v, h)$ with v and h being hyperparameters used to define $[R]$. A connection of this approach to a Tikhonov regularization solution obtained by minimizing the penalized cost

$$C(\vec{\beta}) = (\vec{y} - [X]\vec{\beta})^2 + \lambda \sum_i \left(\frac{\partial^n \beta_i}{\partial i^n}\right)^2, \tag{4.237}$$

where n is the order of the derivatives used for smoothing, is made by Goutte *et al.* Equation (4.237) may be rewritten as

$$C(\vec{\beta}) = (\vec{y} - [X]\vec{\beta})^2 + \vec{\beta}^T[R]\vec{\beta} \tag{4.238}$$

for some $[R]$ (generally different from an $[R]$ derived from a noise autocorrelation model). Equation (4.238) (with the appropriate $[R]$) may be derived as the negative log of the product of the likelihood and prior of Goutte *et al.*'s FIR model with AR noise modeled in the prior. Thus the regularization term in a Tikhonov regularization approach represents a prior covariance when viewed from a Bayesian perspective.

Marrelec *et al.* [299, 300] extend the GLM-based Bayesian model to include several stimuli types per run for multiple runs and show that directed acyclic graphs (DAG) may be used to skip direct calculation of the posterior probability density function (pdf) by representing the posterior distribution as a product of conditional pdfs of the parameters and hyperparameters. The conditional pdfs may be approximated by inverse χ^2 and normal distributions and a Gibbs sampling scheme that is informed by the structure of the DAG may be used to generate a numerical approximation of the posterior distribution.

Woolrich *et al.* [452] represent both temporal and spatial covariance in their Bayesian model but nonlinearly parameterize the IRF, h, used to define the model $h * x$, where x is the stimulus function. They model h as a piecewise continuous addition of four half period cosines in which six parameters are to be determined in MAP solution. The six parameters allow for a characterization of the IRF (different for each voxel) that include the amplitude of the initial dip, the onset delay, the main response amplitude and the undershoot amplitude at the end of the response. MCMC sampling is used to characterize the resulting posterior distribution and a deviance information criterion (DIC) is used to quantify the model goodness of fit. Given a set of parameters θ and data y, the DIC is defined as

$$\text{DIC} = p_D + \overline{D}, \tag{4.239}$$

where $\overline{D} = E_{\theta|y}(D)$ is the posterior expectation of the deviance, D, given by

$$D(\theta) = -2 \ln p(y|\theta) + 2 \ln f(y), \tag{4.240}$$

where f is a nonessential standardizing term, and

$$p_D = \overline{D} - D(E_{\theta|y}(\theta)) \tag{4.241}$$

quantifies the complexity of the model. A low DIC indicates a good model. Woolrich *et al.* find that models that use Markov random fields (MRF) as priors give a low DIC.

A similar nonlinear Bayesian approach is used by Gössl *et al.* [192] but instead of a piecewise function consisting of four half period cosines they use the following piecewise model of the IRF:

$$h(t) = \begin{cases} 0.0 & t < t_1 \\ \exp\left(-\left(\dfrac{t-t_2}{\beta_1}\right)^2\right) & t \in [t_1, t_2) \\ 1.0 & t \in [t_2, t_3) \\ (1+\beta_5)\exp\left(-\left(\dfrac{t-t_3}{\beta_2}\right)^2\right) - \beta_5 & t \in [t_3, t_4) \\ -\beta_5\exp\left(-\left(\dfrac{t-t_4}{4.0}\right)^2\right) & t \geq t_4 \end{cases}, \tag{4.242}$$

where the splice times are defined in terms of the model parameters $\beta_1, \ldots, \beta_5$ by

$$t_1 = \beta_3, \tag{4.243}$$

$$t_2 = \beta_1 + \beta_3, \tag{4.244}$$

$$t_3 = \beta_1 + \beta_3 + \beta_4, \tag{4.245}$$

$$t_4 = \beta_1 + \beta_2 + \beta_3 + \beta_4. \tag{4.246}$$

Note that, with the first segment being zero, the initial dip is not modeled with Equation (4.242).

MRFs are used by Descombes *et al.* [120] at two stages of activation map computation. The first use of MRFs is to denoise the time-series data (see Section 2.4.4). After denoising, a standard SPM is computed and pixels classified by four functions of pixel index s:

(i) The activation map, d_a, such that $d_a(s) \in \{0, -1, 1\}$ with 0 denoting no activation, 1 activation and -1 deactivation (see Section 4.2 for a discussion of deactivation).

(ii) The L^2 norm map, d_n, such that $d_n(s) \in \mathbb{R}^+$.

(iii) The maximum map, d_m, such that $d_m(s) \in \mathbb{R}^+$.

(iv) The time of maximum map, d_i, such that $d_i(s) \in \{1, \cdots, T\}$, where T is the number of time points in the fMRI time-series.

The maps d_n, d_m and d_i are used to define a potential $V^a(s)$ that is used to define MRF that is a "denoised" version of the activation map d_a.

MRFs may be used in a more direct way to compute activation maps by the argument that biological processes should lead to activation patterns that form MRFs. Rajapakse and Piyaratna [367] use this idea and specify a prior condition to constrain the activation map to be an MRF. Smith *et al.* [403] use a similar idea by employing an Ising prior for Bayesian determination of a binary activation map. As an alternative to an MRF prior, Penny and Friston [356] use a prior that specifies that activations should occur as ellipsoid clusters. Penny and Friston's approach belongs to a class of models known as mixtures of general linear models (MGLM).

As an alternative to computationally intensive MCMC methods for computing the posterior distribution, an approximate analytical solution for the posterior density can be obtained by assuming that the approximate posterior density $q(\theta|y)$ can be factored over parameter groups, θ_i, as

$$q(\theta|y) = \prod_i q(\theta_i|y) \tag{4.247}$$

and using the variational Bayes (VB) framework [357]. The VB approach may be understood by first expressing the log of the evidence in terms of an arbitrary distribution $q(\theta|y)$ as

$$\ln p(y) = \int q(\theta|y) \ln p(y) \, d\theta \tag{4.248}$$

$$= \int q(\theta|y) \ln \frac{p(y,\theta)}{p(\theta|y)} \, d\theta \tag{4.249}$$

$$= \int q(\theta|y) \ln \frac{p(y,\theta)}{q(\theta|y)} \, d\theta + \int q(\theta|y) \ln \frac{q(\theta|y)}{p(\theta|y)} \, d\theta \tag{4.250}$$

$$= F + KL, \tag{4.251}$$

then maximizing F via the standard calculus of variations Euler–Lagrange equations leads to the solution for the factors of the approximate posterior

$$q(\theta_i|y) = \frac{\exp[I(\theta_i)]}{\int \exp[I(\theta_i)]} \, d\theta_i, \tag{4.252}$$

where

$$I(\theta_i) = \int q(\theta^{-i}|y) \ln p(y,\theta) \, d\theta^{-i} \tag{4.253}$$

with θ^{-i} being all the parameters not in the ith group. Equation (4.252) may be evaluated analytically. Woolrich *et al.* [451] use VB with an MRF spatial prior.

Rowe [377] describes a multivariate Bayesian model in which a source function s is inferred instead of assuming it to be the given stimulus function x. Both temporal and spatial correlations are modeled using appropriate hyperparameters and an iterated conditional modes (ICM) algorithm (see [377]) is used to find the MAP estimate for s at each voxel.

4.4.2 Multilevel models, empirical Bayes and posterior probability maps

Friston *et al.* [170] outline a general approach to multilevel hierarchical Bayesian modeling in which estimates of variance components at higher levels may be used to provide empirical priors for the variance components at lower levels. This approach is known as empirical Bayes (EB) and the resulting parameter estimate is known as the empirical a-posteriori (EAP) estimate or a parametric empirical Bayesian (PEB) estimate. The use of two hierarchical levels has been applied to two situations. In the first situation the first level is the voxel level and the second level is the subject level [171, 450] to give a mixed-effects model (see Equations (4.119) and (4.120)). In the second situation, the first level is the voxel level and the second level is the spatial level [171, 172]. See Section 5.4 for a third two-level approach where the second level is the epoch in a blocked experimental design. Neumann and Lohmann [338] introduce a hybrid scheme in which the first (voxel) level is a standard SPM analysis and the second level, at the subjects level, is a Bayesian analysis.

A hierarchical Bayesian model with the first level being the voxel level and the second level being the spatial level gives an analysis that replaces the "traditional" SPM approach of computing statistics at the voxel level and then correcting the p values using GRF at the spatial level. This hierarchical Bayesian approach has been implemented in the more recent SPM software[†] and, in place of an SPM, a posterior probability map (PPM) is produced (see Fig. 4.7). Since the SPM software is widely used by neuroscientists, it is worthwhile to examine how the PPM empirical Bayes method works for a single subject following [172]. Begin by partitioning the design matrix and parameter vector into (0) effects of no interest and (1) effects of interest, $[X] = [[X_1][X_0]]$, $\vec{\beta}^T = [\vec{\beta}_1^T \vec{\beta}_0^T]$, and regard the effects of no interest as fixed effects and the effects of interest as random effects. Then the hierarchical observation model is

$$\vec{y} = [[X_1][X_0]] \begin{bmatrix} \vec{\beta}_1 \\ \vec{\beta}_0 \end{bmatrix} + \epsilon^{\vec{(1)}}, \tag{4.254}$$

$$\vec{\beta}_1 = \vec{0} + \epsilon^{\vec{(2)}}, \tag{4.255}$$

where the effects of interest are estimated by considering the whole volume in which they are posited to have a mean of 0 and a covariance structure $E(\|\epsilon^{\vec{(2)}}\|^2) = \sum_{i=1}^m \lambda_i [E_i]$. Here $m = \dim \vec{\beta}_1$, λ_i is a hyperparameter and $[E_i]$ is the "basis matrix"

[†] SPM2 is available at http://www.fil.ion.ucl.ac.uk/spm.

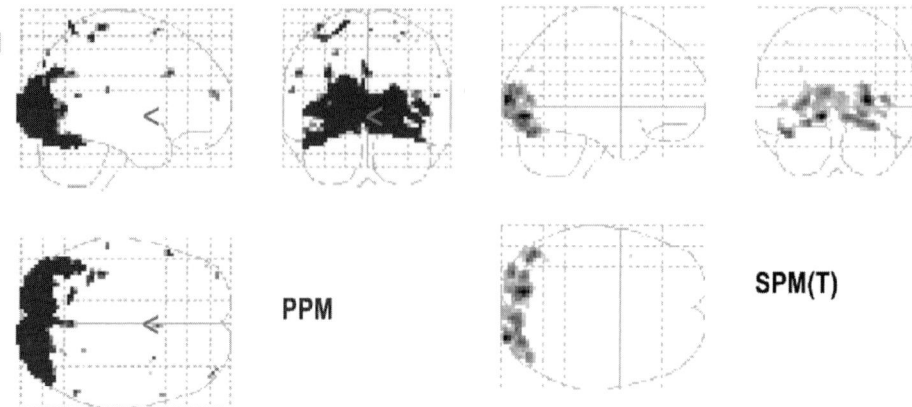

Fig. 4.7. Comparison of a PPM activation map with an SPM *t* map. The PPM
shows voxels that are activated with a 95% probability or greater. The SPM *t* map
shows voxel level activations, corrected using GRF (see Section 4.1.8), at $p < 0.05$.
Note the increased sensitivity of the PPM. The images are shown as displayed in
the output of the SPM software on the MNI space (see Section 2.5) "glass brain".
Taken from [172], used with permission.

with 1 on the *i*th diagonal and zeros elsewhere. Note that it is assumed that there is
no spatial correlation in this particular model but it could be added. The model of
Equations (4.254) and (4.255) may be expressed as a single-level model by stuffing
the random effects into a compound error term:

$$\vec{y} = [X_0]\vec{\beta}_0 + \vec{\xi}, \qquad (4.256)$$

$$\vec{\xi} = [X_1]\vec{\epsilon}^{(2)} + \vec{\epsilon}^{(1)}. \qquad (4.257)$$

Here $E_{\text{all voxels}}(\vec{\xi}\vec{\xi}^T) = [C_\xi] = [Q]\vec{\lambda}$ in which

$$[Q] = [[X_1][E_1][X_1]^T, \cdots, [X_1][E_m][X_1]^T, [V]] \qquad (4.258)$$

and

$$\vec{\lambda}^T = [\lambda_1^T \cdots \lambda_m^T \lambda_{\epsilon_p}^T], \qquad (4.259)$$

where $[V]$ gives a model of the time-series autocorrelation and the hyperparameter
λ_{ϵ_p} is a pooled estimate of voxel level hyperparameter λ_ϵ (see below). Let $[Y]$ be the
complete data matrix (in which voxel time courses appear as rows, one row for each
voxel), let $[C_{\epsilon_p}] = \lambda_{\epsilon_p}[V]$. Then the values for variance hyperparameters $\vec{\lambda}$ may be
estimated using the EM algorithm in which the E-step consists of (starting with a
guess for $\vec{\lambda}$) computing the *expectation* values for $[C_\xi]$ and a posterior covariance
matrix $[C_{\vec{\beta}_0|\vec{y}}]$ via

$$[C_\xi] = [Q]\vec{\lambda} \qquad (4.260)$$

and

$$[C_{\vec{\beta}_0|\vec{y}}] = ([X_0]^T [C_{\epsilon_p}]^{-1} [X_0])^{-1}. \tag{4.261}$$

The values of $[C_{\xi}]$ and $[C_{\vec{\beta}_0|\vec{y}}]$ from the E-step are then used to update (*maximize*) a scoring function, represented by $[H]$ and $\vec{g}$, in the M-step by computing

$$[P] = [C_{\xi}]^{-1} - [C_{\xi}]^{-1} [X_0][C_{\vec{\beta}_0|\vec{y}}][X_0]^T [C_{\xi}]^{-1}, \tag{4.262}$$

$$g_i = -\frac{1}{2}\mathrm{tr}([P][Q_i]) + \frac{1}{2}\mathrm{tr}([P]^T [Q_i][P][Y]^T [Y]/n), \tag{4.263}$$

$$H_{ij} = \frac{1}{2}\mathrm{tr}([P][Q_i][P][Q_j]), \tag{4.264}$$

where n is the number of voxels in the data set. With $[H]$ and $\vec{g}$, the old value of $\vec{\lambda}$ may be updated via

$$\vec{\lambda}_{\mathrm{new}} = \vec{\lambda}_{\mathrm{old}} + [H]^{-1}\vec{g} \tag{4.265}$$

and the new $\vec{\lambda}$ is used in the E-step in repetition until $\vec{\lambda}$ converges. The value of $\vec{\lambda}$ found is maximized within the null space of $[X_0]$ so it is an ReML estimate. The estimate of $\vec{\lambda}$ may then be used to construct the EB prior $[C_\beta]$ for the original voxel level model:

$$[C_\beta] = \begin{bmatrix} \sum_{i=1}^m \lambda_i[E_i] & [0] \\ [0] & [\infty] \end{bmatrix}, \tag{4.266}$$

where $[\infty]$ is a $\dim(\vec{\beta}_0)$ square matrix with infinities on the diagonal and zeros elsewhere to represent the flat priors assumed for the effects of no interest. The variance components $[C_\epsilon]$ and $[C_{\vec{\beta}|\vec{y}}]$ for the original voxel level model given by

$$\vec{y} = [X]\vec{\beta} + \vec{\epsilon} \tag{4.267}$$

may be estimated (ReML estimates) using another EM process to estimate the hyperparameter λ_ϵ, where the E-step consists of computing

$$[C_\epsilon] = \lambda_\epsilon[V], \tag{4.268}$$

$$[C_{\vec{\beta}|\vec{y}}] = ([X]^T [C_\epsilon]^{-1} [X] + [C_\beta]^{-1})^{-1}, \tag{4.269}$$

and the M-step consists of computing

$$[P_\epsilon] = [C_\epsilon]^{-1} - [C_\epsilon]^{-1} [X_0][C_{\vec{\beta}|\vec{y}}][X_0]^T [C_\epsilon]^{-1}, \tag{4.270}$$

$$g_\epsilon = -\frac{1}{2}\mathrm{tr}([P_\epsilon][V]) + \frac{1}{2}\mathrm{tr}([P_\epsilon]^T [V][P_\epsilon]\vec{y}\vec{y}^T), \tag{4.271}$$

$$H_\epsilon = \frac{1}{2}\mathrm{tr}([P_\epsilon][V][P_\epsilon][V]). \tag{4.272}$$

After each M-step, λ_ϵ is updated via

$$\vec{\lambda}_{\epsilon,\mathrm{new}} = \vec{\lambda}_{\epsilon,\mathrm{old}} + H_\epsilon^{-1} g_\epsilon \tag{4.273}$$

until it converges. Finally the posterior parameter estimates are given by

$$\vec{\eta}_{\vec{\beta}|\vec{y}} = [C_{\vec{\beta}|\vec{y}}][X]^T[C_\epsilon]^{-1}\vec{y} \tag{4.274}$$

and the posterior probability that a contrast of parameters exceeds a threshold γ is given by

$$p = 1 - \Phi\left(\frac{\gamma - \vec{c}^T\vec{\eta}_{\vec{\beta}|\vec{y}}}{\sqrt{\vec{c}^T[C_{\vec{\beta}|\vec{y}}]\vec{c}}}\right), \tag{4.275}$$

where $\vec{c}$ is the contrast vector and Φ is the unit normal cumulative probability density function.

4.4.3 Random walk models

Gössl *et al.* [191] construct a locally linear state space model in which the drift and activation parameters are modeled as random walks. With $y_{i,t}$ being the brightness of voxel i at time point t and $z_{i,t}$ being the value at time t of the model HRF, constructed in the usual way by convolving an IRF model with the stimulus time course, their model may be expressed as

$$y_{i,t} = a_{i,t} + z_{i,t}b_i + \epsilon_{i,t}, \qquad \epsilon_{i,t} \sim N(0, \sigma_i^2), \tag{4.276}$$

$$a_{i,t} = 2a_{i,t-1} - a_{i,t-2} + \xi_{i,t}, \quad \xi_{i,t} \sim N(0, \sigma_{\xi i}^2), \tag{4.277}$$

$$b_{i,t} = 2b_{i,t-1} - b_{i,t-2} + \eta_{i,t}, \quad \eta_{i,t} \sim N(0, \sigma_{\eta i}^2), \tag{4.278}$$

where $a_{i,t}$ models, in a random walk fashion, the drift and $b_{i,t}$ models the activation. The imposition of a Gaussian prior on the "innovations" $\xi_{i,t}$ and $\eta_{i,t}$ constrains the walks to be locally linear by penalizing deviations from straight lines. A consequence of the random walk model is that an activation map can be computed for every time point, giving an activation movie. Frames from such a movie are shown in Fig. 4.8. Also the model for the HRF is specific for each voxel i, being parametrized by parameters θ_i and d_i (dispersion and delay) as

$$z_{i,t} = \sum_{s=0}^{t-d_i} h(s, \theta_i)x_{t-d_i-s}, \tag{4.279}$$

where h is the model IRF and x is the stimulus function. Parameter estimation proceeds in two steps with θ_i and d_i being estimated in the first step by minimizing the distance between z_i and y_i using a Gauss–Newton method in the first step and a Kalman filter and smoother embedded in an EM algorithm (in lieu of a computationally intense MCMC method) in the second step to estimate a and b. In [193] Gössl *et al.* introduce spatial as well as temporal correlations into their

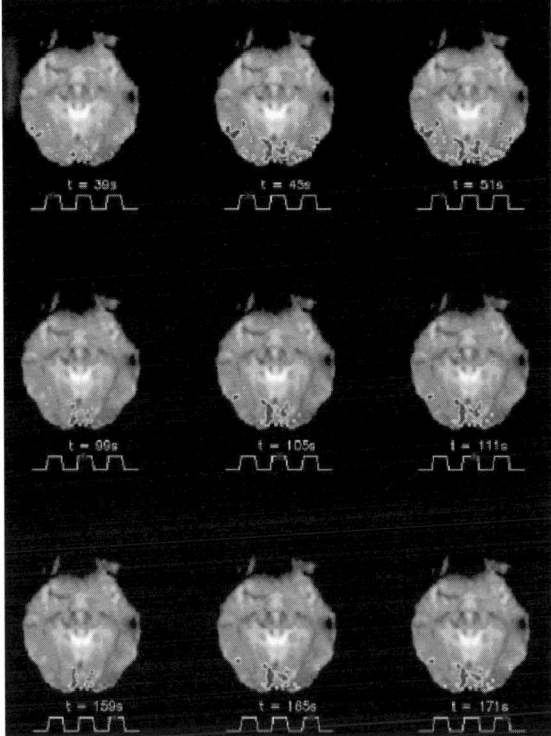

Fig. 4.8. Example frames from an activation movie made with the random walk model of Equations (4.276)–(4.278) and a visual stimulation paradigm. Although hard to see in this reproduction, underneath the images is a representation of the presentation paradigm and a red dot on that representation showing the time of the activation map. The times, from top left to bottom right are $t = 39, 45, 51, 99, 105, 111, 159, 165$ and 171 s. Taken from [191], used with permission. See also color plate.

model through the use of a global hyperparameter, λ_i, for each time-point to characterize the spatial correlations. This global hyperparameter approach makes the multivariate approach possible where an ML solution of a full multivariate GLM would not be.

4.5 Nonparametric methods

One of the more popular nonparametric statistics that has been used for the computation of activation maps for blocked designs is the Kolmogorov–Smirnov (KS) statistic[†] that measures the maximum distance between two cumulative probability

---

[†] The lyngby software, available at http://hendrix.imm.dtu.dk/software/lyngby/ is capable of producing KS statistic base activation maps [207].

distributions. If x_i represents the brightness of a voxel at time-point i in a time-series that contains N_{on} task points and N_{off} rest time-points then $S_{N_{on}}(x)$ represents the cumulative distribution of the frequencies of x_i values for i in the task set. The cumulative distribution of the rest set $S_{N_{off}}(x)$ is defined similarly and the KS statistic, D, is given by

$$D = \max_{-\infty < x < \infty} |S_{N_{on}}(x) - S_{N_{off}}(x)|. \tag{4.280}$$

The p value associated with D for rejecting H_0 that $D = 0$ is

$$p = Q_{KS}(\lambda) = 2 \sum_{k=1}^{\infty} (-1)^{k-1} e^{-2k^2\lambda^2}, \tag{4.281}$$

where

$$\lambda = \left(\sqrt{N_e} + 0.12 + 0.11/\sqrt{N_e} \right) \tag{4.282}$$

and

$$N_e = \frac{N_{on}N_{off}}{N_{on} + N_{off}}. \tag{4.283}$$

When used over all the voxels, the p value needs to be Bonferroni corrected by multiplying by the number of comparisons (voxels) considered. Aguirre *et al.* [5] measure false positive rates (FPR) on null task data sets that are thresholded using D values that correspond to $\alpha = 0.05$ (given by setting $p = \alpha$ in Equation (4.281) and solving for D) for individual images and for a Bonferroni corrected α across images. They find significantly higher false positive rates than α using the signed rank test and a binomial test to look for differences between α and the FPR. They test the KS statistic derived activation maps from raw data, linear drift corrected data and high-pass filtered data (see Section 4.1.1) and find that the linear drift corrected data lead to slightly lower FPR while high-pass filtered data had considerably higher FPR than maps made from raw data. Aguirre *et al.* also test the normality of the null data using the D'Agostino–Pearson K^2 statistic that measures the skewness and kurtosis of the sample and find only a slight deviation from normality. Since the KS statistic is a less powerful statistical test when the data are normally distributed and because of the high observed FPR, Aguirre *et al.* advise that activation maps based on the KS statistic are less desirable than maps based on standard SPM approaches.

The other popular nonparametric approach to activation map computation is to use permutation methods to empirically describe the null distribution of a statistic of interest [341]. The essentials of permutation methods have been covered in Section 4.1.7, where it was shown that Fourier or wavelet methods are needed to permute the data because of the autocorrelation in the time-series.

4.6 Fourier methods

Fourier methods for computing activation maps are limited to periodic blocked designs. One approach is to detect a sinusoidal signal at the block frequency [365]. For that method, first compute the periodogram, $I(\omega_k)$, of the voxel time-series of length N

$$I(\omega_k) = \frac{1}{N} \left| \sum_{n=0}^{N-1} x_n \exp(-i\omega_k n) \right|^2, \quad \text{where } \omega_k = 2\pi k/N \quad (4.284)$$

and then compare

$$F = \frac{I(\omega_p)}{(1/(N-3)) \left[\sum_{k=0}^{N-1} I(\omega_k) - I(0) - 2I(\omega_p) \right]} \quad (4.285)$$

to $F_{2,N-3}$, where ω_p (a multiple of $2\pi/N$) is the frequency of the block presentation.

Kiviniemi et al. [255] use the fast Fourier transform (FFT) to compute a spectrum for every voxel and also, for the global signal, to look for BOLD activity in resting brains. A frequency of interest is identified in the global spectrum and the mean and standard deviation of the amplitude of that frequency in the spectrums of all the brain voxels is computed. Voxels whose intensity at the frequency of interest are more than six standard deviations above the mean are considered activated. Kiviniemi et al. compare this FFT approach to correlation, PCA and ICA (see Section 5.2) methods for identifying activation in the resting brain[†]. The FFT method finds the smallest volume of activation and the correlation method finds the largest volume.

Mitra and Pesaran [321] describe how windowed Fourier analysis may be applied to a singular value decomposition (SVD) eigenvector set (see Section 5.1) of the voxel time courses to construct a time-frequency analysis in which quasi-periodic physiological processes (heart beat and respiration) may be identified along with the task correlated activity. This method requires T_R to be short enough to adequately sample the physiological processes. Mitra and Pesaran also describe a "space-frequency" SVD in which SVD analysis is applied to the fMRI data set in which all the time-series have been replaced with windowed Fourier transforms. From the space-frequency SVD, a frequency–coherence graph may be computed in which the coherence reflects how much of the fluctuation in the windowed frequency band is captured by the dominant SVD spatial eigenimage.

[†] Kiviniemi et al. [255] use the FastICA software, available at http://www.cis.hut.fi/projects/ica/fastica/ to do their ICA analysis.

Hansen *et al.* [209] describe how to analyze blocked designs using a GLM in which columns $2k - 1$ and $2k$ are given by the sinusoids

$$x_{n,2k-1} = \sin(k\omega_0 t_n), \tag{4.286}$$

$$x_{n,2k} = \cos(k\omega_0 t_n) \tag{4.287}$$

for $k \in \{1, \dots, K\}$ and ω_0 is the fundamental frequency of the BOLD signal. Hansen *et al.* regard K and ω_0 as parameters and show how the posterior probability $p(\omega_0, K|\vec{y})$ may be computed using the principle of conjugate priors.

4.7 Repeatability and comparisons of methods

As should be obvious, there are many ways to compute an activation map with advantages and disadvantages to each method, not the least of which is ease of use. No one approach can reveal the best or true activation map, especially since the physiology of the BOLD signal is still under active investigation (see Section 4.2). Reviewed in this section are studies that address not only the consistency of maps computed using different methods but also the consistency of maps computed using the same method and experimental paradigm but for different sessions and/or subjects.

The identification of what comprises a "valid" activation map depends in the first instance on converging evidence on the phenomenon being investigated and what "makes sense" to the neuroscientist [265]. So the first step after map computation is to visually inspect the computed maps [359] both in terms of whether the map "makes sense" and, especially in the case of blocked designs, in terms of if the time course shows a smooth BOLD response to the tasks [389].

Beyond subjective comparisons, the simple measure of reproducibility, R_v/M, where R_v is the number of times a voxel appears activated in M replicates, has been used. Employing this measure of reproducibility, Yetlin *et al.* [465] find an optimal threshold for the simple correlation approach ($r = 0.60, p = 5 \times 10^{-8}$ uncorrected) and Tegeler *et al.* [419] conclude that a simple *t*-test gives better reproducibility than multivariate discriminant analysis (see Section 5.1) at 4 T. Both investigators use the finger-touching task. Lukic *et al.* [290] find, using an ROC analysis[†], the opposite result that multivariate discriminant analysis produces better results than a simple *t*-test. Using a nonparametric Friedman two-way ANOVA, Machulda *et al.* [294] find that varying the threshold level (correlation in this case) and maximum cluster size analysis parameters (factor 1) produces relative changes in activation between four ROIs (posterior and anterior hippocampus, parahippocampal gyrus

---

[†] Lukic *et al.* [290] use the LABROC1 software for the ROC analysis, which may be obtained from http://xray.bsd.uchicago.edu/krl/KRL_ROC/software_index.htm.

and entorhinal cortex – factor 2). Using ROC analysis (applied to resting brain ROIs with artificial activations added), Skudlarski *et al.* [401] come to the following conclusions about optimal approaches:

- *Preprocessing*: The removal of drifts and high pass filtering is good, while temporal normalization, smoothing and low pass filtering are not beneficial.
- *Map statistics*: The tested statistics, t, r and the nonparametric Mann–Whitney statistic, all give similar results.
- *Task design*: Within the blocked designs tried, a task on, task off of 18 s each is found to be optimal.
- *Spatial clustering*: Smoothing before map computation is found to be more efficient than cluster filtering (removing small clusters) after map computation.

Skudlarski *et al.* take the approach that the statistic threshold is not to be interpreted probabilistically; the statistic is only a metric to be used for the detection of activation. Below we will see how an optimal threshold may be selected on the basis of replicated experiments.

McGonigle *et al.* [301] look at intersession effects in a study involving one subject, three tasks and 99 sessions (33 sessions/task) and find significant session-by-task interaction. However, the voxels showing the session-by-task interaction do not show up in multisession fixed-effects analysis. This indicates that voxels found active in a single session for one subject may be due to some effect of the session and not the given task.

Saad *et al.* [382] examine the effect that averaging scans (time-series) together for a single subject has on the spatial extent of the detected activation. The scans are averaged by transforming data to Tailarach space before averaging (see Section 2.5). They find that the spatial extent of the activations increases monotonically without asymptote when the number of scans averaged together is increased from 1 to 22.

Lange *et al.* [267] suggest that several different methods be used for every study in order to validate the final reported map. They look at nine methods, comparing the results using: (i) ROC analysis, (ii) a concordance correlation coefficient, ρ_c, for reproducibility between two random variables with means μ_1 and μ_2 and variances σ_1^2 and σ_2^2 defined as

$$\rho_c = \frac{2\rho\sigma_1\sigma_2}{\sigma_1^2 + \sigma_2^2 + (\mu_1 - \mu_2)^2}, \qquad (4.288)$$

where ρ is the correlation between the two random variables, and (iii) a resemblance of activity measure based on the similarity of spatial correlations in the two activation maps.

When several methods are used to produce summary activation maps, one can also average the summary maps to yield a consensus map and avoid having to pick the "best" map [208].

Genovese *et al.* [179] present a probabilistically based quantitative method for assessing the reliability of maps produced over M replications of an identical experiment with a fixed analysis method. Using a model that consists of a linear combination of binomial distributions for the probability that a truly active voxel is classified as active, p_A, (true positive) and for the probability that a truly inactive voxel is classified as active, p_I, (false positive) Genovese *et al.* construct a log-likelihood function to characterize p_A, p_I and λ, the proportion of truly active voxels, as

$$\ell(p_A, p_I, \lambda | \vec{n}) = \sum_{k=0}^{M} n_k \ln[\lambda p_A^k (1 - p_A)^{(M-k)} + (1 - \lambda) p_I^k (1 - p_I)^{(M-k)}],$$

$$(4.289)$$

where $\vec{n} = [n_1, \ldots, n_M]^T$ and n_k is the number of voxels in replicate k that are classified as active. Maximizing ℓ produces the estimates $\hat{p}_A, \hat{p}_I$ and $\hat{\lambda}$ with standard errors given by the diagonal elements of the Hessian (matrix of second derivatives) of ℓ. When comparing two map computation methods one can use two log-likelihood functions of the form of Equation (4.289), one for each method, and constrain λ to be the same for both log-likelihood functions. However, analysis of the differences in p_A between two methods must take into account the correlation between the two p_A values when the variance of the differences is computed. Maximizing the log-likelihood requires that $M \geq 3$ in order for a solution to exist (a condition known as identifiability). If $M = 3$, an ROC can be constructed to define a relationship between p_A and p_I so that Equation (4.289) may then be maximized. To address the reliability issue with Equation (4.289), compute the *ML reliability efficient frontier*, which is a curve of the relationship between $\hat{p}_A$ and $\hat{p}_I$ as a function of the tuning parameter (which is usually the SPM threshold). With that relationship, the reliability criterion function $p_O = \lambda p_A + (1 - \lambda)(1 - p_I)$ may be maximized to find the optimal threshold setting (see Section 4.1.8). Applying these methods, Noll *et al.* [343] find nearly identical between-session reliability for a motor and a working memory task. Liou *et al.* [279] show how p_O, which defines the observed proportion of agreement between the true active/inactive status and the classification result, may be corrected for chance. Defining the agreement expected by chance as $p_C = \lambda \tau + (1 - \lambda)(1 - \tau)$, where $\tau = \lambda p_A + (1 - \lambda) p_I$, and using an ROC model, the proportion of agreement corrected for chance is

$$\rho = \frac{p_O - p_C}{1 - p_C}. \qquad (4.290)$$

For k different contrasts (or computation methods) the agreement between maps from contrasts j and k may be assessed with

$$\kappa_{ij} = \frac{P_j P_k - \sum_i n_{ij} n_{ik} / [M(M-1)V]}{P_j P_k}, \qquad (4.291)$$

where n_{ij} is the number of times that the ith voxel is considered active for the jth contrast out of M replications and P_j is the sum of $n_{ij}/(MV)$ over the V voxels in the data set. Then optimum thresholds for the k contrasts may be found by maximizing

$$\kappa = \frac{\sum_{j \neq k} P_j P_k \kappa_{jk}}{\sum_{j \neq k} P_j P_k}. \tag{4.292}$$

Once optimal thresholds are found the quantity, R_v/M, where R_v is the number of replications for which the voxel was labeled active, may be plotted for each voxel to give a *reproducibility map*. Liou *et al.* suggest that most truly active voxels have a reproducibility above 90% (strongly reproducible in their terms), while some have a reproducibility between 70% and 90% (moderately reproducible). Liou *et al.* apply these methods to find optimal points on the ROC for an EM method and a GLM method and find improved sensitivity without an increase in the false alarm rate for the EM method.

4.8 Real time fMRI and complementary methods

Converging evidence for hypotheses regarding brain activation and function at the neural level may be obtained by combining fMRI data with other neurological data. These other approaches include the use of EEG [220] and single-neuron electrical recording [117]. Efforts are also underway to detect the neural magnetic field directly using MRI [50, 51] with some success being reported by Xiong *et al.* [463]. The approach to obtaining such magnetic source MRI (msMRI) data is very similar to conventional fMRI methods except that the experimental paradigm is designed so that the BOLD signal is held at a constant level to allow the detection due to spin phase changes caused by neuronal magnetic fields.

For application in the clinical setting, activation maps need to be computed in real time or near real time so that physicians can be sure that adequate data have been obtained before the patient leaves the MRI scanner. To implement real time fMRI, fast computational algorithms have been developed for real time image coregistration [110], recursive correlation coefficient computation [108] and GLM parameter computation through Gram–Schmidt orthogonalization of the design matrix [25]. Prototype real time fMRI hardware systems capable of the multiple required tasks including task/imaging synchronization, physiological and task (e.g. button presses) monitoring and map computation have been implemented [404, 434].

5

Multivariate approaches: connectivity maps

The description of connectivity between brain regions may be usefully given by two concepts, those of functional connectivity and effective connectivity. *Functional connectivity* refers to temporal correlations between spatially separated brain regions, while *effective connectivity* refers to a causal relationship between spatially separated brain regions [147]. Functional connectivity may be assessed by the variance observed between a pattern vector (volume image) $\vec{p}$ and the data $[M]$. Here we adopt the convention of organizing fMRI time-series data into a $v \times n$ data matrix $[M]$, where v is the number of voxels in a volume (scan) and n is the number of time-points in the fMRI time-series. In other words, the entire volume for one scan is represented as a row in $[M]$ and each row represents a different scan time. We also assume that the grand mean vector has been subtracted from the data vectors in $[M]$ so that the variance–covariance matrix of the data is given simply by $[C] = [M]^T[M]$. Then, with this setup, if the quantity $\|[M]\vec{p}\|^2$ is large it provides evidence for a functional connectivity pattern that matches $\vec{p}$. Each component of the vector $[M]\vec{p}$ is the covariance (unnormalized correlation) between a scan volume and $\vec{p}$ (see Equation (1.9)), so if $\|[M]\vec{p}\|^2$ is large it means that the correlations vary considerably in time. Ways of quantifying functional connectivity are reviewed in Section 5.1. Ways of quantifying effective connectivity are reviewed in Section 5.6.

Many authors cite their favorite introductory multivariate statistics texts because the ideas of multivariate statistics need to be relatively well understood before applications to fMRI may be understood. My favorite introductory texts are those by Rencher [371, 372].

5.1 The GLM – multivariate approaches

In general, the multivariate analysis of fMRI time-series data is problematical because the dimension, v, of the data vector, represented by the number of 3D voxels in a volume data set, far exceeds the number of data vectors measured, represented by the number of time points, n. One way around this is to restrict analysis to a predetermined small number of ROIs [342, 370]. The ROIs represent anatomically homogeneous regions and the time-series of the voxels in each ROI may be averaged together to reduce the noise component. In this way the dimension of the data vector is reduced so that $v < n$ and a multivariate analysis, which explicitly models the covariances between the data vector components, may be used.

Another way to reduce the dimension of the data vector is to use principal component analysis (PCA, see Section 5.2) which may be derived from an SVD of the data matrix $[M]$ [156]. The SVD of $[M]$ is given by

$$[M] = [U][S][V]^T, \tag{5.1}$$

where, if $[M]$ is an $n \times v$ matrix, $[U]$ is an $n \times n$ matrix, $[S]$ is an $n \times n$ diagonal matrix of (square roots of) eigenvalues and $[V]$ is a $v \times n$ matrix. The columns of $[V]$ are the (orthonormal) eigenvectors of the sums of squares and products matrix $[C] = [M]^T[M]$; these vectors represent the eigenimages (and which may be displayed as images) that account for the amount of variance in the data as quantified by the corresponding eigenvalue (in $[S]$). The columns of $[U]$ contain the voxel time courses that correspond to the eigenimages in $[V]$. The columns of $[U]$ are the (orthonormal) eigenvectors of $[M][M]^T$ because

$$[M][M]^T = [U][S]^2[U]^T, \tag{5.2}$$

as may be easily derived from Equation (5.1). Since the matrix $[C]$ is unmanageably large, the eigenimage matrix $[V]$ (the principal components of $[M]$) may be computed from the eigenvalue solution $[U]$ using

$$[V] = [M]^T[U][S]^{-1}. \tag{5.3}$$

From the SVD, we may produce a reduced set of time courses to analyze, using standard multivariate techniques like MANOVA, MANCOVA and discriminant analysis, as

$$[X] = [U][S], \tag{5.4}$$

where, generally, only eigenvectors with eigenvalues greater than 1 are kept in $[X]$. The reduced representation of the fMRI time course data in $[X]$ is now amenable to analysis by a multivariate GLM

$$[X] = [G][\beta] + [\epsilon], \tag{5.5}$$

where $[G]$ is the design matrix, $[\beta]$ is the parameter matrix and $[\epsilon]$ is the error matrix. Typically $[G]$ represents a MANCOVA model[†] such that $[G] = [H|D]$, where the submatrix $[H]$ models effects of interest and $[D]$ models the effects (covariates) of

---

[†] MANCOVA = Multivariate ANalysis of COVAriance.

no interest. The model for $[G]$ used by Friston *et al.* [156] contains four columns for each of three conditions for a total of twelve columns with the four columns for each condition containing Fourier basis functions (cosines) for modeling the HRF for the given condition. The least squares estimate of the parameters is given by

$$[\hat{\beta}] = ([G]^T[G])^{-1}[G]^T[X] \tag{5.6}$$

with variance (assuming iid Gaussian errors)

$$\text{var}(\vec{\beta}_j)[a^T|g^T] = \sigma_j([G]^T[G])^{-1}, \tag{5.7}$$

where $\vec{\beta}_j$ is the *j*th column of $[\hat{\beta}]$ and $[a]^T$ represents the parameters of interest and $[g]^T$ the parameters of no interest. The omnibus statistical significance of the parameters may be tested using Wilk's Lambda

$$\Lambda = \frac{\det[R(\Omega)]}{\det[R(\Omega_0)]}, \tag{5.8}$$

where $[R(\Omega)]$ is the sums of squares and products matrix due to the error,

$$[R(\Omega)] = ([X] - [G][\hat{\beta}])^T([X] - [G][\hat{\beta}]), \tag{5.9}$$

and $[R(\Omega_0)]$ is the sums of squares and products matrix due to the error after discounting the effects due to the hypotheses of interest $[H]$,

$$[R(\Omega_0)] = ([X] - [D][g])^T([X] - [D][g]). \tag{5.10}$$

The statistic $-(r - ((J - h + 1)/2)\ln(\Lambda)$ may be compared to a χ^2 distribution with *Jh* degrees of freedom to test the null hypothesis, where $r = n - \text{rank}([G])$, *J* is the number of columns in $[X]$ and $h = \text{rank}([H])$.

The effects of interest from the MANCOVA of Equation (5.5) are given by

$$[T] = ([H][a])^T([H][a]), \tag{5.11}$$

which may be used to interpret the eigenimages (functional connectivity) in terms of the MANCOVA model in a canonical variates analysis (CVA). CVA produces a sequence of contrasts $\vec{c}_i$ such that the variance ratio

$$\frac{\vec{c}_i^T[T]\vec{c}_i}{\vec{c}_i^T[R(\Omega)]\vec{c}_i} \tag{5.12}$$

is maximized subject to the constraint that $\langle \vec{c}_i, \vec{c}_j \rangle = 0$ for $j < i$ (there is no constraint for $\vec{c}_1$). The solution of the canonical contrast vectors is given by the solution $[c]$ to the eigenvalue problem

$$[T][c] = [R(\Omega)][c][\theta], \tag{5.13}$$

where $[\theta]$ is a diagonal matrix of eigenvalues (canonical values) and the columns of $[c]$ are the canonical contrast vectors. The canonical (functional connectivity)

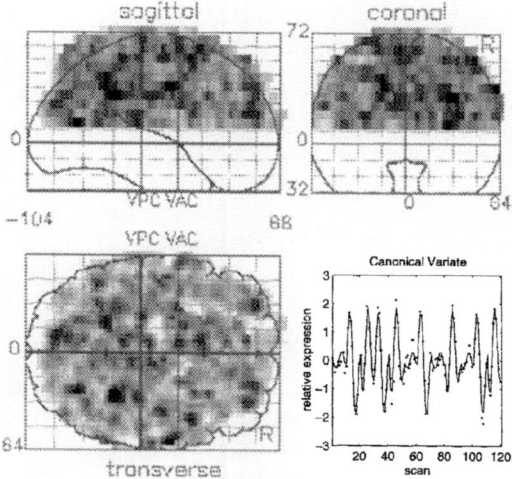

Fig. 5.1. An example of an functional connectivity image, $\vec{C}_1$, from a CVA as given by Equation (5.14) for a three-condition task, where the tasks were a rest condition, moving left and right hands in a fixed alternating order in response to a visual cue and, moving left and right hands in a random alternating order in response to a visual cue. The dotted time course in the lower right graph is $[X]\vec{c}_1$, the first canonical variate, where $[X]$ is the reduced set of time courses as given by Equation (5.4), and the solid line is $[G]\vec{c}_1$, where $[G]$ is the design matrix. Taken from [156], used with permission.

images (see Fig. 5.1) are given by the columns of $[C]$, where

$$[C] = [V][c]. \tag{5.14}$$

The statistical significance of the canonical functional connectivity images $\vec{C}_j$ for $j \leq S + 1$ may be tested by comparing the statistic $(r - ((J - h + 1)/2)) \ln] [\prod_{i=S+1}^{J}(1 + \theta_i)]$ to a χ^2 statistic with $(J - s)(h - S)$ degrees of freedom. These null distributions assume that the reduced data, in the form of components, follow multivariate normal distributions which, in general, is an approximation. Almeida and Ledberg [10] give exact distributions to use with component models when the data are normally distributed. Nandy and Cordes [336] outline ROC methods that may be used to assess the sensitivity and specificity of multivariate methods. A classical application of SVD and CVA to both the complete data set and to subsequently identified brain ROIs is given by Bullmore *et al.* [68].

5.1.1 Anatomically constrained analysis

Another way of reducing the dimension of the scan observation vector is to restrict the analysis from a 3D volume to a 2D manifold that represents the cortical surface

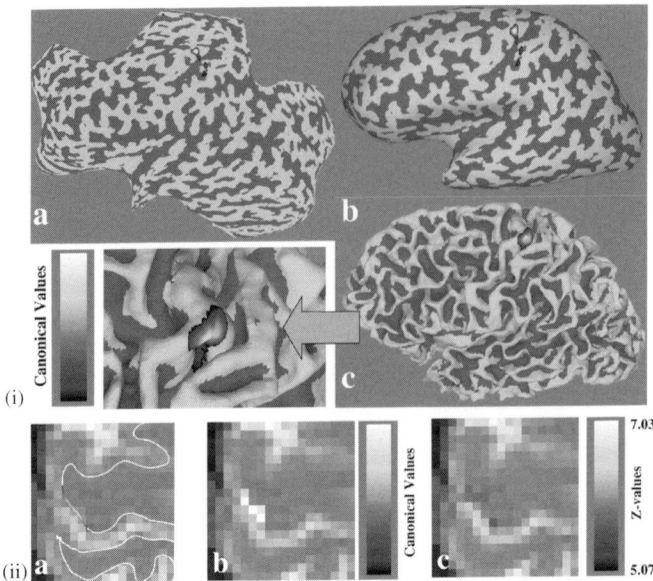

Fig. 5.2. Activations computed on the 2D surface of the cortex using AIBFs. In part (ii): (a) shows an outline of the computed cortical surface on top of the voxels of the original EPI data along with the activation as computed on the 2D surface; (b) shows the activations as transformed by the function g of Equation (5.15) from cortical space to voxel space; (c) shows a conventional map as computed with the SPM software for comparison. Part (i) shows the same activation on: (a) the cortical flat map S_F; (b) the inflated cortical map S_I; and (c) the folded grey matter surface S_G. Taken from [247], used with permission. See also color plate.

of the brain, see Fig. 5.2. Kiebel *et al.* [247] use the Brain Voyager software (Brain Innovations, Rainer Goebel) to (i) segment the gray matter from the rest of the brain to give a gray matter surface S_G, (ii) inflate S_G to produce a balloon like surface S_I and (iii) flatten S_I, after cutting, onto a flat 2D surface, S_F. Circular Gaussian basis functions $\{b_F^j | 1 \leq j \leq N_p\}$ are defined on S_F and transformed back to S_G, which is defined by a set of vertices $\vec{V}_G$ and faces, to provide a set of basis functions $\{b_G^j\}$. Then a function f on S_G may be represented by $f = \sum_j c_j b_G^j$ and, if g represents an appropriate transformation from $L^2(S_G)$ to $L^2(\mathbb{R}^3)$ then[†] the data $\vec{Y}$ from one scan may be modeled with

$$\vec{Y} = g(f(\vec{V}_G)) + \vec{\epsilon}, \tag{5.15}$$

where $f(\vec{V}_G)$ represents the vector formed by evaluating f at each of the components of $\vec{V}_G$. Let N_K represent the number of brain voxels in a scan volume and let

---

[†] I use L^2 for purposes of illustration and not for any rigorous mathematical purposes. The map g in Equation (5.15) of course involves a restriction to segmented gray matter.

[A] be the $N_K \times N_p$ matrix in which the jth column represents the basis vector $b_Y^j = g(b_G^j)$ (anatomically informed basis functions – AIBFs) in voxel coordinates. Then Equation (5.15) may be written as

$$\vec{Y} = [A]\vec{\beta} + \vec{\epsilon}, \qquad (5.16)$$

which may be solved using

$$\vec{\hat{\beta}} = ([A]^T[A] + \lambda[I])^{-1}[A]^T\vec{Y}, \qquad (5.17)$$

where the regularization parameter can be chosen to be

$$\lambda = \frac{\mathrm{tr}([A]^T[A])}{N_p}. \qquad (5.18)$$

With Equations (5.16) and (5.17) we may convert the measured voxel data $\vec{Y}_j$ at time j to measured cortical surface data $\vec{\hat{\beta}}_j$ and so obtain a time-series of cortical surface data. The cortical data matrix is then $[B]^T = [\vec{\hat{\beta}}_1 \cdots \vec{\hat{\beta}}_{N_Y}]$, where N_Y is the number of time-points (scans). The cortical data matrix may also be expressed in voxel space as $[B]^T_{\mathrm{voxel}} = [A][B]^T$. Either way, the resulting cortical data set may be analyzed using the univariate methods of Chapter 4 or by the multivariate methods of Section 5.1 with [B] taking the place of [M]. For multisubject studies, the cortical surface in a standard stereotaxic space (see Section 2.5) may be used as S_G in place of the individual segmented cortical surface [248]. That is, the data are smoothed and transformed to the standard stereotaxic space where the basis functions b_G^j are defined. The smoothing requires that the spatial GLM of Equation (5.16) be multiplied by a smoothing matrix $[L_E]$ before $\vec{\hat{\beta}}$ is computed, similarly to how temporal smoothing, through $[K]$, is used in Equation (4.25).

Smoothing is also possible on the cortical surface at the subject level using an analogy between Gaussian convolution on a surface and heat diffusion on a surface [14]. Through the analogy a smoothed version of the data may be computed on the irregular cortical surface vertex grid by evolving a finite difference diffusion model to a predetermined time to reflect convolution with a Gaussian kernel with a predefined width.

5.2 Principal and independent component methods

PCA typically gives the eigenvector component time courses specified by the columns of the matrix [U] in Equation (5.1). Spatial versions of PCA give eigen-images, which are the columns of [V] of Equation (5.1). The PCA components are ordered, by the magnitude of their corresponding eigenvalues, as the directions in data space in which the data vary maximally. The PCA eigenvectors define the hyperellipsoid axes of the data scatterplot if the data are multivariate normally distributed. PCA is often invoked as a method of reducing the dimensionality of subsequent analysis such as independent component analysis or clustering. Normally, PCA is done on the whole data set, with perhaps extracerebral voxels trimmed to reduce the computational burden. Lai and Fang [263] partition the fMRI time-series into three sets with one set being the middles (to avoid transitions) of the active time blocks and one set being the middles of the inactive time blocks, concatenate the active and inactive segments and perform a PCA of the new data set. The active and inactive segments for each voxel are then projected into the subspace spanned by the M most dominant principal components to form $\vec{f}_{(a)}$ and $\vec{f}_{(i)}$, respectively, at each voxel. Then they define activation, A, at a given voxel by

$$A = (\|\vec{f}_{(a)}\| - \|\vec{f}_{(i)}\|) \sum_{j=1}^{M} \frac{|m_{(a),j} - m_{(i),j}|}{\sqrt{\sigma^2_{(a),j} - \sigma^2_{(i),j}}}, \qquad (5.19)$$

where $m_{(a),j}$ and $\sigma_{(a),j}$ represent the mean and standard deviation of the projection of $\vec{f}_{(a)}$ into principal component j in the active section (with similar definitions for the inactive segment), and $\| \cdot \|$ represents the ℓ^2 norm (the active and inactive segments have equal length). Thresholding A gives the active voxels.

A standard application of PCA, in general, is to use the components to define factors that may be associated with some process. The use of PCA for factor analysis has been applied to fMRI by Backfrieder *et al.* [24] and by Andersen *et al.* [11]. Assume that there are m time-series images and p dominant principal components (say the ones with eigenvalues greater than 1). Then we may model the $v \times m$ (transpose of the) data matrix $[Y]$ (with $m > p$) as

$$[Y] = [V][B] + [E], \qquad (5.20)$$

where $[V]$ is the $v \times p$ matrix containing the p most dominant principal components as columns, $[B]$ is a $p \times m$ matrix that allows the original data set to be expressed as linear combinations of the dominant principal components and $[E]$ is the residual matrix. The basis of the signal space specified in the columns of $[V]$ may be replaced by another, not necessarily orthogonal, basis represented by columns in a matrix $[F]$ that represent the factor directions. Specifically, there is a transformation $[T]$ such that

$$[V][T][T]^{-1}[B] = [F][C], \qquad (5.21)$$

so that

$$[Y] = [F][C] + [E], \qquad (5.22)$$

where $[C]$ is a matrix that specifies the "loadings" of the signal onto the factor components and each column of $[F]$ specifies a factor image. If $[T]$ is an orthogonal transformation, then it is said to perform an "orthogonal rotation" of the principal components into the factors[†], otherwise the transformation is said to be an "oblique rotation". A variety of methods exist to find the appropriate $[T]$ (e.g. Varimax) but usually some a-priori information is needed. For example, when a known activation region is identified, $[T]$ can be chosen so that the active region appears in only one or two factor images representing activation and gray matter (anatomical) factors.

If the data are not multivariate normally distributed, then the maximum variation may not be along one direction in data space but along a curve. In that case, which will happen when there are interactions among sources (factors), a nonlinear PCA analysis may be used. Friston *et al.* [165, 169] describe a second-order nonlinear PCA. Following Friston *et al.* let $\vec{y}(t) \in \mathbb{R}^v$ represent the fMRI data set consisting of v voxels when t is restricted to p values. Nonlinear PCA assumes that the data are a nonlinear function, $\vec{f}$, of J underlying sources represented by $\vec{s}(t) \in \mathbb{R}^J$ for each t as

$$\vec{y}(t) = \vec{f}(\vec{s}(t)). \tag{5.23}$$

Truncating the Taylor expansion of Equation (5.23) after the bilinear terms gives

$$\vec{y}(t) = \vec{V}^{(0)} + \sum_{j=1}^{J} u_j(t)\vec{V}_j^{(1)} + \sum_{j=1}^{J}\sum_{k=1}^{J} u_j(t)u_k(t)\vec{V}_{j,k}^{(2)}, \tag{5.24}$$

where $\vec{u}(t) = \vec{s}(t) - \bar{\bar{s}}$ and

$$\vec{V}^{(0)} = \vec{f}(\bar{\bar{s}}), \tag{5.25}$$

$$\vec{V}_j^{(1)} = \left[\frac{\partial f_1}{\partial u_j}(\bar{\bar{s}}), \cdots, \frac{\partial f_v}{\partial u_j}(\bar{\bar{s}}) \right]^T, \tag{5.26}$$

$$\vec{V}_{j,k}^{(2)} = \left[\frac{\partial^2 f_1}{\partial u_j \partial u_k}(\bar{\bar{s}}), \cdots, \frac{\partial^2 f_v}{\partial u_j \partial u_k}(\bar{\bar{s}}) \right]^T. \tag{5.27}$$

Equation (5.24) may be generalized by adding a sigmoid function σ to allow for more general interactions:

$$\vec{y}(t) = \vec{V}^{(0)} + \sum_{j=1}^{J} u_j(t)\vec{V}_j^{(1)} + \sum_{j=1}^{J}\sum_{k=1}^{J} \sigma(u_j(t)u_k(t))\vec{V}_{j,k}^{(2)}. \tag{5.28}$$

Given $\vec{u}$, it is possible to find $\vec{V}^{(0)}$, $\vec{V}_j^{(1)}$ and $\vec{V}_{j,k}^{(2)}$ for $1 \le j, k \le J$ by minimizing the residuals (difference between the left and right hand sides of Equation (5.28)) in a least squares sense. Friston *et al.* use neural network methods to find the

---

[†] Mathematically, all rotations are orthogonal but language is not always logical. Here oblique and orthogonal refer to the relative rotations of the basis vectors.

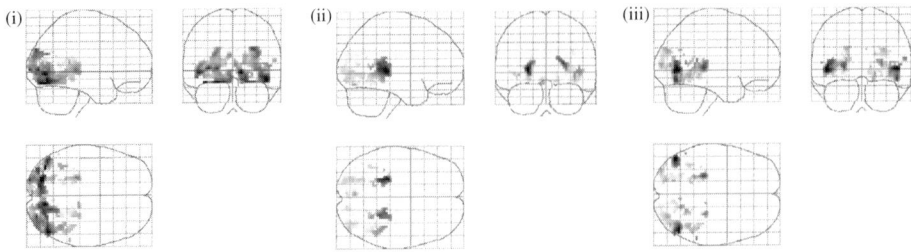

Fig. 5.3. An example result from a nonlinear PCA. The analysis gives, in this case, (i) spatial mode 1, $\vec{V}_1^{(1)}$, (ii) spatial mode 2, $\vec{V}_2^{(1)}$, and (iii) their second order interaction mode, $\vec{V}_{1,2}^{(2)}$. Taken from [169], used with permission.

sources $\vec{u}$ that minimize the residuals globally. The end result is J sources, $u_i(t)$, $1 \le i \le J$, whose amplitudes are characterized by $\vec{V}_j^{(1)}$ and whose interactions are characterized by $\vec{V}_{j,k}^{(2)}$, see Fig. 5.3.

5.2.1 Independent component analysis (ICA)

From a multivariate perspective, an fMRI data set may be thought of in two ways. In the usual way (as we have been implicitly assuming until now) the data are represented by a $p \times v$ data matrix $[M]$, in which each row represents one 3D voxel data set taken at a fixed scan time. That way, $[M]$ represents p measurements of a v-dimensional quantity, the spatial image. The other way is to represent the data by a $v \times p$ matrix $[Y] = [M]^T$, in which each column represents one 3D voxel data set taken at a fixed scan time. Viewed this way we have v measurements of a p-dimensional quantity, the voxel intensity time course. An ICA of $[M]$ produces spatially independent components (component images) and is termed spatial ICA (sICA). An ICA of $[Y]$ produces temporally independent components (component timecourses) and is termed temporal ICA (tICA) [79]. More specifically, let $[\hat{W}]^{-1}$ be the estimate of the mixing matrix and $[C]$ be the matrix containing the components as rows. The sICA approach gives

$$[M] = [\hat{W}]^{-1}[C] \tag{5.29}$$

and the columns of $[\hat{W}]^{-1}$ give the time courses corresponding to the spatial components in the rows of $[C]$. In general the number of rows in $[C]$ is chosen to be less than the number of rows in $[M]$. The tICA approach, which generally is used only on an ROI basis to limit the number of computations required, gives

$$[Y] = [\hat{W}]^{-1}[C] \tag{5.30}$$

and the columns of $[\hat{W}]^{-1}$ give the spatial images corresponding to the time course components in rows of $[C]$.

The premise behind ICA is that the data are considered to be mixtures of sources, represented in the rows of $[S]$ according to

$$[M] = [A][S], \tag{5.31}$$

where $[A]$ is the mixing matrix. The goal is to find the unmixing matrix $[W] = [A]^{-1}$ so that

$$[S] = [W][M]. \tag{5.32}$$

ICA will produce estimates $[C]$ and $[\hat{W}]$ of $[S]$ and $[W]$ respectively to give

$$[C] = [\hat{W}][M]. \tag{5.33}$$

The sources are assumed to be statistically independent in the sense that

$$p(S_1, \ldots, S_n) = \prod_{k=1}^{n} p(S_k) \tag{5.34}$$

when n components (maximally n equals the number of time points for sICA) are used [303, 415], where S_i is the ith signal source, or component, represented in the ith row of $[S]$. Such statistical independence is a stronger condition of independence than orthogonality, as assumed in PCA. There are several measures of independence but one of the most common is mutual information. Working now with the estimated components[†] C_1 to C_n the mutual information is defined as

$$I(C_1, \ldots, C_n) = \sum_{i=1}^{n} H(C_i) - H(C_1, \ldots, C_n), \tag{5.35}$$

where

$$H(x) = - \int p(x) \, \ln p(x) \, dx \tag{5.36}$$

is the entropy of the probability distribution p associated with x. The infomax ICA method maximizes the mutual information by iterating estimates for $[W]$ in a neural net training algorithm. Esposito *et al.* [130] compare the infomax algorithm to a fixed-point algorithm and find that infomax is superior for finding noise components while the fixed-point algorithm gives more accurate active components when assessed by correlation to a model waveform.

PCA has been compared to ICA both in terms of its ability to find noise components [421], so that the data may be denoised by projecting into the subspace

---

[†] I have temporarily dropped the ⁻ notation because the component vectors are row vectors.

orthogonal to the noise, and in terms of separating signals from two different tasks performed in the same run [49]. ICA was found to be superior to PCA on both counts. However, McKeown and Sejnowski [304] examine the assumptions that ICA produces a small number of independent components mixed with Gaussian noise and find that those assumptions are more likely to be true in white matter than in active regions containing blood vessels. They therefore suggest that ICA is more suited to ROI analysis than to global analysis of fMRI data.

The use of PCA before ICA can reduce the dimensionality of the ICA analysis and reduce the computational burden, especially for very long (>1000 time-points) time-series [306]. In terms of probability distribution moments, PCA separates components to second order while ICA separates components to all orders. McKeown *et al.* [303] illustrate an intermediate, fourth order, method of component separation due to Comon [101].

ICA has been used for the analysis of functional connectivity in the brain both in a resting state [17, 431] and in standard task driven investigations and to compare resting data to task driven data [17]. But in general, it can be difficult to interpret the components produced by ICA [87]. So ICA frequently needs to be used in combination with other methods. One approach is to correlate a model HRF with the component time courses to determine which components are task related [324]. Intravoxel correlations of the original time-series within selected sICA component maps may also be used [254]. ICA can also be combined with clustering methods (see Section 5.4). Himberg *et al.* [219] generate several ICAs of a data set, for subsequent clustering purposes, by bootstrapping the data and/or beginning the optimization (maximization of mutual information) from different points[†]. (Bootstrapping is an alternative to data permutation, where from a set of *n* data values a new set of *n* data values is generated by randomly picking a data value from the original sample, recording it and replacing the data point back in the original data set and then repeating the random selection and replacement until a "bootstrapped" sample of *n* values is taken from the original sample.) McKeown [305] introduces a hybrid technique, HYBICA, that results in a method intermediate between a completely data driven ICA and a completely model driven GLM. McKeown uses ICA to define the signal subspace and the columns of interest in the GLM design matrix. First, *k* task-related independent components are identified to define an activation subspace. Then the stimulus function, *u*, is projected into the basis vectors of the *k*-dimensional activation subspace to define *k* columns of interest in the design matrix. The dimension *k* can be varied from 1 to *n*, the length of the time-series, to vary the GLM from more data driven to more model driven.

[†] Himberg *et al.*'s software, Icasso, is available from http://www.cis.hut.fi/projects/ica/icasso. They also use the FastICA and SOM MATLAB toolboxes available from http://www.cis.hut.fi/research/software.html.

ICA may be applied to the data set consisting of Fourier transforms, or power spectrums, of each voxel time-series. Such an approach is particularly useful for blocked designs where the frequency of presentation can be readily identified to define the task-related components [325]. ICA of the Fourier spectrum also has the advantage of being insensitive to variation in the delay of the response, produc-ing components that are dependent on time course shape independently of delay [83]. ICA of complex-valued fMRI data is also possible with the advantage of potentially separating responses due to large and small blood vessels into different components [82].

To use ICA with groups of subjects, two approaches have been proposed. One is to concatenate data across subjects after dimension reduction with PCA for each subject [80]. With the second approach, components are computed for each indi-vidual subject, a component of interest (e.g. one containing the motor region) is selected from one subject and the components from each of the other subjects that correlate (spatially) the most with the original component are selected [81]. An alternative is to use, as the original map of interest, the sum of two compo-nents, examining its correlations with two component sums in the other subjects. Using two components combines a potential activation component with a poten-tial anatomical component (e.g. a gray matter component). A standard fixed- or mixed-effects analysis, or a conjunction analysis, may then be done on the selected components (see Section 4.1.9).

5.2.2 Canonical correlation analysis

Canonical correlation analysis, introduced to fMRI by Friman *et al.* [145, 146], represents another way, besides PCA and ICA, to decompose multivariate data into uncorrelated components. With PCA the variance, or signal energy, is maximized in each component in turn. With ICA, Gaussianity-related measures like kurtosis, negentropy and/or mutual information are minimized in each component in turn to produce mutually statistically independent components. With canonical correla-tion analysis, the autocorrelation of the resulting components is maximized with the idea that the most interesting components have the most autocorrelation. As with PCA and ICA we may compute either a temporal canonical correlation analysis to find temporal components or a spatial canonical correlation analysis to find spatial components. Following Friman *et al.* we give the details for a temporal canonical correlation analysis using the time course $x_i(t)$ in $\vec{x}(t) = [x_1(t), \ldots, x_n(t)]^T$, where n is the number of pixels (for a slice by slice analysis) and $t \in \{1, \ldots, N\}$, as the data vector for each pixel measurement. The spatial canonical correlation analysis proceeds similarly with the data vector being the scan (image) for each time meas-urement. The idea is that the time courses are a linear combination, or mixing, of some underlying sources: $\vec{x}(t) = [A]\vec{s}(t)$, where $[A]$ is the mixing matrix. Canonical correlation analysis finds the matrix $[W]$ such that $\hat{\vec{s}}(t) = [W]\vec{x}(t)$. Defining $\vec{y}(t) = \vec{x}(t-1)$, to maximize the lag 1 autocorrelation, the rows of $[W]$ are the eigenvectors that solve the eigenvalue problem

$$[C_{xx}]^{-1}[C_{xy}][C_{yy}]^{-1}[C_{yx}]\vec{w} = \rho^2\vec{w}, \tag{5.37}$$

where the estimates for the correlation matrices are given by

$$[\hat{C}_{xx}] = \frac{1}{N}\sum_{t=1}^{N}\vec{x}(t)\,\vec{x}(t)^T, \tag{5.38}$$

$$[\hat{C}_{yy}] = \frac{1}{N}\sum_{t=1}^{N}\vec{y}(t)\,\vec{y}(t)^T, \tag{5.39}$$

$$[\hat{C}_{xy}] = [\hat{C}_{yx}]^T = \frac{1}{N}\sum_{t=1}^{N}\vec{x}(t)\,\vec{y}(t)^T \tag{5.40}$$

and the square root of the eigenvalues, ρ, represents the correlation between $\vec{w}^T\vec{x}(t)$ and $\vec{w}^T\vec{y}(t) = \vec{w}^T\vec{x}(t-1)$ with the first eigenvector and value giving the largest possible correlation. The second eigenvalue is the largest possible correlation in the subspace orthogonal to the first eigenvector, etc. The components $s_i(t)$ are maximally autocorrelated signal sources that should contain the activation time course(s) of interest (typically the second or third component after the drifts). Once the components of interest are identified (see Fig. 5.4), correlations of the voxel time courses with those components give activation and/or functional connectivity maps. To reduce computation, a subset of the PCA components may be used as the data for both ICA and canonical correlation analysis.

5.2.3 Functional data analysis (FDA)

As discussed in Section 2.4.4, FDA represents the fMRI data set as a smooth function. Lange *et al.* [267] apply FDA to produce a regularized PCA. They use a cubic B-spline basis of 42 basis functions, $B_p(t)$, $1 \le p \le 42$, representing the data with the coefficient matrix $[C]$ (coefficients of the basis functions in place of the original voxel values, i.e. $[C]$ replaces $[M]$). To use $[C]$ they let $[J]$ be the matrix of inner products $\langle B_p, B_q\rangle$, $[K]$ be the matrix of inner products $\langle D^2B_p, D^2B_q\rangle$ and define $[L]$ by the Cholesky decomposition $[L][L]^T = [J] + \lambda[K]$, where λ is a roughness penalty parameter. Next $[L][D] = [C]$ is solved for new coefficients $[D]$ and a PCA performed on $[D]$ to find eigenvectors $[U]$. The equation $[L]^T[Z] = [U]$ is solved for $[Z]$ and the columns of $[Z]$, $\vec{z}_p$, renormalized so that $\vec{z}_p^T[J]\vec{z}_p = 1$.

PCA   ICA   CCA

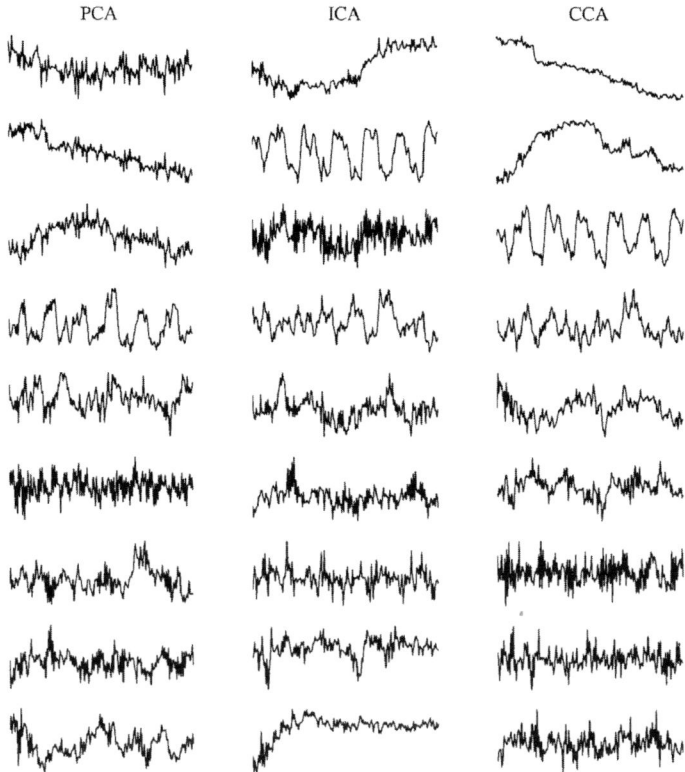

Fig. 5.4. Time courses extracted from fMRI data by temporal PCA, ICA and canonical correlation analysis. Here the PCA components were used, to reduce the dimensionality of the subsequent processing, as inputs to the ICA and canonical correlation analysis. The statistically independent ICA components emerge from the analysis in no particular order which can make the identification of the activation time course difficult. The canonical correlation analysis components emerge in order of those with the highest to lowest autocorrelation. Here the first component is clearly drift and the third component is clearly the activation component. Correlating the third canonical correlation analysis component with the original data then gives an activation map. Taken from [145], used with permission.

Finally, the (function) eigenvectors $\zeta = [Z]^T[B]$ are computed. Lange *et al.* find that the second eigenvector represented the (unthresholded) activation map.

5.2.4 Multivariate assessment of design matrices

PCA may be obtained from the SVD of the data matrix $[M]$ as given by Equation (5.1). Given the design matrix $[X]$ of a multivariate GLM of the form

$$[M] = [X][\beta] + [\epsilon], \tag{5.41}$$

more meaningful, and interpretable, SVD decompositions are possible. The methods are based on decomposing the sums of squares and products matrix $[M]^T[M]$ into model-based sums of products and squares $[H]$ and error-based sums of products and squares $[R]$ as follows

$$[M]^T[M] = ([X][\hat{\beta}] + [\epsilon])^T([X][\hat{\beta}] + [\epsilon]) \tag{5.42}$$

$$= [\hat{\beta}][X]^T[X][\hat{\beta}] + [\epsilon]^T[\epsilon] \tag{5.43}$$

$$= [H] + [R], \tag{5.44}$$

where the least squares estimate of $[H]$ is given by

$$[H] = [M]^T[X]([X]^T[X])^{-1}[X]^T[M]. \tag{5.45}$$

Following Kherif *et al.* [245] we list three of these more general SVD approaches.

- *Partial least squares* [302] is based on an SVD of $[X]^T[Y]$, giving eigenimages obtained from optimal combinations of predictors in $[X]$ from the point of view of explaining variance in terms of those predictors. A more direct decomposition of $[H]$ is given by an orthonormalized partial least squares which is an SVD of $([X]^T[X])^{-1/2}[X]^T[M]$.
- *Canonical variates analysis* seeks the matrix of images (columns) $[V]$ that maximizes $([V][H][V]^T)/([V][R][V]^T)$ thus maximizing the SNR of the resulting components (see Section 5.1). Finding that $[V]$ can be done through the SVD of the matrix

$$[Z] = ([X]^T[X])^{-1/2}[X]^T[M][R]^{-1/2} \tag{5.46}$$

or, more directly, through the SVD of

$$[Z] = ([X]^T[X])^{-1/2}[X]^T[M]([M]^T[M])^{-1/2} \tag{5.47}$$

using the relation

$$\max_{[V]} \frac{[V]([M]^T[M])[V]^T}{[V][R][V]^T} = \max_{[V]} \frac{[V][H][V]^T}{[V][R][V]^T} + 1. \tag{5.48}$$

- *Multivariate linear model* (MLM) is a variation of CVA that takes into account the temporal autocorrelation noise matrix $[\Sigma]$ and results from an SVD of

$$[Z] = ([X]^T[\Sigma][X])^{-1/2}[X]^T[M][N]^{-1/2}, \tag{5.49}$$

where

$$[N]^{-1/2} = \mathrm{diag}(1/\hat{\sigma}_1, 1/\hat{\sigma}_2, \ldots) \tag{5.50}$$

$$= \mathrm{diag}((\vec{\hat{\epsilon}}_1^T\vec{\hat{\epsilon}}_1/v)^{-1/2}, (\vec{\hat{\epsilon}}_2^T\vec{\hat{\epsilon}}_2/v)^{-1/2}, \ldots), \tag{5.51}$$

where v is the number of voxels. Kherif *et al.* show how the MLM method may be used with large design matrices and a test data set to narrow the choice of a GLM model for the analysis of a new experimental data set.

5.2.5 Parallel factor (PARAFAC) models

PARAFAC models, introduced to fMRI by Andersen and Rayens [12], represent data using more than two dimensions with each dimension corresponding to a factor. For example, for a trilinear model, if the factors are space (voxels), time and subject, then the data are represented by a data *cube*, S, where the third dimension indexes subject number. For a data set composed of I voxels, J time points and K subjects, a PARAFAC model has R components of the form

$$S = \sum_{r=1}^{R} c_r \vec{x}_r \otimes \vec{y}_r \otimes \vec{z}_r, \tag{5.52}$$

where $\otimes$ denotes a tensor or outer product and $\vec{x}_r \in \mathbb{R}^I$, $\vec{y}_r \in \mathbb{R}^J$ and $\vec{z}_r \in \mathbb{R}^K$ are unit vectors. The vectors $\vec{x}_r$, $\vec{y}_r$ and $\vec{z}_r$ define the spatial (functional connectivity), temporal and subject components respectively. PARAFAC models may be solved using eigenbased methods or by an iterative alternating least squares routine.

5.3 Wavelet methods

The wavelet transform has many properties that make it attractive for application to fMRI analysis [70]. One of those properties is that, when the wavelet function has enough vanishing moments, the wavelet transform provides a Karhunen–Loéve like decomposition of a voxel time-series. That is, the wavelet coefficients are uncorrelated (see Section 4.1.7). Noise with $1/f$ spectral properties may be described as distributional derivatives of fractional Brownian motion whose fractal or self-similar properties may be characterized with the Hurst exponent $0 < H < 1$. In slightly more precise terms, for $H \in (0, 1)$ there exists exactly one Gaussian process that is the stationary increment of the corresponding self-similar process. When $H \in (0, \frac{1}{2})$, the process has short term dependence, when $H \in (\frac{1}{2}, 1)$ the process has long term dependence and $H = \frac{1}{2}$ corresponds to white noise. The spectral density, S, of a fractional Brownian increment process is

$$S(f) \approx \frac{\sigma_H^2}{f^\gamma}, \tag{5.53}$$

where

$$\sigma_H^2 = \sigma^2 (2\pi)^{-2H} \sin(\pi H) \Gamma(2H + 1) \tag{5.54}$$

and the exponent is $\gamma = 2H - 1$. Fadili and Bullmore [136], in a method they call wavelet-generalized least squares (WLS), build a likelihood function based on the correlation properties of the wavelet coefficients of a $1/f$ autocorrelated time-series (specifically for autoregressive fractionally integrated moving average (ARFIMA) noise with fractional difference parameter $d = H - \frac{1}{2}$) to enable an ML estimate of the correlation matrix $[\Sigma_J]$ in the wavelet domain (it will be diagonal)[†]. This estimate may be used with the wavelet transform of a GLM of the time-series to compute the BLUE of the GLM parameters. Specifically the GLM

$$\vec{y} = [X]\vec{\beta} + \vec{\epsilon} \qquad (5.55)$$

is transformed under the discrete wavelet transform (DWT) to yield

$$\vec{y}_w = [X_w]\vec{\beta} + \vec{\epsilon}_w \qquad (5.56)$$

and the BLUE of $\vec{\beta}$ is given by

$$\hat{\vec{\beta}} = ([X_w]^T[\hat{\Sigma}]^{-1}[X_w])^{-1}[X_w]^T[\hat{\Sigma}]^{-1}\vec{y}_w. \qquad (5.57)$$

One of the earlier applications of wavelets to activation map computation applied the wavelet transform twice, once spatially for denoising and smoothing as described in Section 2.4.5 and then temporally to identify significant clusters (within a detail scale level j of interest) of correlated wavelet coefficients [55, 218]. The spatial transform increases the contrast to noise level of BOLD signals whose spatial extent is comparable to the level of smoothing [467]. Using the estimate σ_j as given by Equation (2.57) for the temporal wavelet transform, the active voxels are found by first computing

$$B_j(i/n) = \frac{1}{\sigma_j\sqrt{2n}} \sum_{k=1}^{i}(d_{j,k}^2 - \bar{d}_j^2), \qquad (5.58)$$

where $\bar{d}_j$ is the average of the detail coefficients at level j, and the KS statistic

$$K_j = \max_{1 \leq i \leq n} |B_j(i/n)|. \qquad (5.59)$$

Voxels with significant K_j are then considered active.

Ruttimann *et al.* [380] consider a blocked experimental design with task on and off conditions and analyze the 3D data set given by the mean over the blocks of the difference of mean on and mean off signals. That is, if $g_i^{(1)}(\vec{p})$ represents the mean voxel values for the on-block number i and $g_i^{(0)}(\vec{p})$ represents the mean voxel values for the off-block number i, then the data set

$$f(\vec{p}) = \frac{1}{N} \sum_{i=1}^{N}(g_i^{(1)}(\vec{p}) - g_i^{(0)}(\vec{p})) \qquad (5.60)$$

is analyzed. Müller *et al.* [332] generalize $f(\vec{p})$ to represent a contrast from a standard fMRI GLM. Activation map computation then proceeds in a two-step

---

[†] Knowledge of an ARFIMA model of $[\Sigma_J]$ at each voxel also allows one to plot H maps.

elimination of wavelet coefficients (setting them to zero) of the wavelet transform of f, followed by an inverse wavelet transform of the surviving wavelet coefficients to produce the activation map. The first culling step consists of eliminating wavelet coefficients by saving only those coefficients, $d_{j,k}^m$, whose value of $(d_{j,k}^m/\sigma_N)^2$ lies in the tail of an appropriate χ^2 distribution, where σ_N is a noise standard deviation estimated from the data[†]. The second step consists of rejecting the remaining coefficients that fall below a threshold on the assumption that they follow a normal distribution. Desco *et al.* [118] investigate the effect of using different wavelet functions for this method.

Meyer [316] begins with a partially linear model for a voxel time-series

$$y_i = \theta_i + \beta x_i + v_i, \tag{5.61}$$

where the subscript i denotes time or, equivalently, the component of the implied vectors. The vector $\vec{\theta}$ is a nonparametric model for the trend, $\vec{x}$ denotes a single column design matrix (for simplicity) and $\vec{v}$ represents noise that can be modeled as a $1/f$ noise process. The trend is modeled as a linear combination of large scale wavelets at a scale coarser than a given J_0. If the wavelet has p vanishing moments, a polynomial trend of degree $p - 1$ can be approximated with small error. The smaller J_0 is, the more the trend will follow the given time-series. The model of Equation (5.61) wavelet transformed to

$$[W]\vec{y} = [W]\vec{\theta} + \beta[W]\vec{x} + [W]\vec{v}, \tag{5.62}$$

where the covariance matrix of the noise, $[W]\vec{v}$, will be given by

$$[\Sigma] = \mathrm{diag}(\sigma_J^2, \sigma_{J-1}^2, \ldots, \sigma_1^2), \tag{5.63}$$

where σ_i are as given by Equation (2.57). Concatenating $[W]\vec{\theta}$ with β, after eliminating the zero coefficients of $[W]\vec{\theta}$, modifies Equation (5.62) to

$$[W]\vec{y} = [A]\vec{\xi} + [W]\vec{v}, \tag{5.64}$$

which has an ML solution of

$$\vec{\hat{\xi}} = ([A]^T[\Sigma]^{-1}[A])^{-1}[A]^T[\Sigma]^{-1}[W]\vec{y}, \tag{5.65}$$

where $\vec{\hat{\xi}}$ contains the wavelet coefficients of $\vec{\theta}$ and the parameter β.

5.3.1 Continuous wavelet transforms

Continuous wavelet transforms may also be used to detect activation in a method von Tscharner *et al.* [424] label WAVANA. The continuous wavelet transform of

---

[†] The m superscript denotes direction, for example in Fig. 2.5 there are three directions, horizontal, vertical and diagonal.

a time-series $S(t)$ gives a function $S_w(j, t)$ in which the scale level j and time t may both be regarded as continuous variables. The function $S_w(j, t)$ at fixed j is a band-pass version of the original time-series and characterizes differences between scales in a continuous manner. The function $S_w(j, t)$ may be displayed as a contour plot in the (j, t) plane (this plot is known as a scalogram). For a blocked design of N cycles, the function $S_w(j, t)$ may be partitioned into N functions $S_{wp}(j, n, \tau)$, where τ is the time from the beginning of each cycle. An average function

$$S_a(j, \tau) = \frac{1}{N} \sum_{n=1}^{N} S_{wp}(j, n, \tau) \tag{5.66}$$

and a standard deviation function

$$\sigma_{S_a}(j, \tau) = \sqrt{\frac{1}{N} \sum_{i=1}^{N} [S_{wp}(j, n, \tau) - S_a(j, \tau)]^2} \tag{5.67}$$

may be defined. Then two t-statistics functions may be computed

$$S_t(j, n, \tau) = \frac{S_{wp}(j, n, \tau)}{\sigma_{S_a}(j, \tau)}, \tag{5.68}$$

$$S_{at}(j, \tau) = \frac{S_a(j, \tau)}{\sigma_{S_a}(j, \tau)/\sqrt{N}} \tag{5.69}$$

giving, potentially, an activation map for every time t in the time-series or for every cycle time τ but in practice the maximums of the two functions are used to produce maps. Within the contour plot of $S_w(j, t)$ the scale j that best represents the periodic paradigm may be identified (usually $j = 1$) and activation maps computed for that scale.

In another application of the continuous wavelet transform, Müller *et al.* [333] used the complex-valued Morlet wavelet

$$\psi(t) = e^{i\omega_0 t} e^{-t^2/2} \tag{5.70}$$

to compute the phase of the resulting continuous cross wavelet transform between two functions g and f given by

$$W_{a,b}^{\psi}(f \times g) = W_{a,b}^{\psi}(f) \, W_{a,b}^{\psi}(g)^*, \tag{5.71}$$

where g is a reference function and f is a voxel (or ROI averaged) time series. The inverse variance of $W_{a,b}^{\psi}(f \times g)$ shows if f and g have coherence and may be used to infer phase differences between active brain regions (for inferring mental chronometry) or to define active regions because lower strength activations have higher phase variability.

Shimizu *et al.* [397] apply the continuous wavelet transform at each voxel, using the Mexican hat wavelet, to compute a multifractal spectrum. The multifractal spectrum is a function $D(h)$ of the Hölder coefficient h whose value is the Hausdorff dimension of underlying fractal processes in the signal specific to the Hölder coefficient. The Hölder coefficient may be defined at each time point t_0 in the time-series, f, and is defined as the largest exponent h such that there exists a polynomial $P_n(t)$ of order n that satisfies

$$|f(t) - P_n(t - t_0)| = \mathcal{O}(|t - t_0|^h) \tag{5.72}$$

for t in a neighborhood of t_0. Shimizu *et al.* show how properties of the "maxima skeleton" of the continuous wavelet transform may be used to compute the function $D(h)$, which has a simple "inverted U" shape from which one may extract the point (h_{max}, D_{max}), and w, its FWHM. The quantity

$$P_c = \frac{h_{max}}{D_{max}} w \tag{5.73}$$

may then be used to define the activation map.

5.4 Clustering methods

Clustering consists of assigning each voxel time course vector $\vec{y}_j \in \mathbb{R}^P$ to one of K clusters, $C_k \subset \{1, \ldots, N\}$, on the basis of minimizing the within-class (intraclass) inertia

$$I_W = \frac{1}{N} \sum_{k=1}^{K} \sum_{j \in C_k} d^2(\vec{y}_j, \vec{c}_k), \tag{5.74}$$

where N is the number of voxels in the data set, $\vec{c}_k$ is the class center and d is a distance measure on $\mathbb{R}^P$, and on the basis of maximizing the between-class (interclass) inertia

$$I_B = \frac{1}{N} \sum_{k=1}^{K} |C_k| \, d^2(\vec{c}_k, \bar{\bar{c}}), \tag{5.75}$$

where $|C_k|$ is the cardinality of C_k and $\bar{\bar{c}} = \sum_{k=1}^{K} (|C_k|/N)\vec{c}_k$ [194]. If the class centers are defined to be the average of the vectors in the cluster, then the minimization of I_W becomes equivalent to the maximization of I_B via Huygens's familiar formula that states that the total sum of squares is equal to the within sum of squares plus the between sum of squares. Thus, minimizing Equation (5.74) leads to a definite assignment of each voxel time course to a definite cluster. A fuzzy probabilistic assignment is possible by minimizing Equation (5.87) instead. Clusters are computed iteratively by assigning vectors to the cluster of the nearest center

then recalculating the cluster centers and reassigning vectors to clusters, until convergence. This K-means method requires only that the number of clusters, K, be
specified in advance. To avoid having to choose K, or to help in choosing K, a
number of *hierarchical clustering* schemes have been proposed.

The group-average agglomerative hierarchical method [194] begins with the
maximum number of clusters, one per data vector. Next the two closest vec-
tors/clusters are joined, reducing the number of clusters by 1. The process proceeds
until there is one cluster. Then some method is required to pick, from the hierarchy,
the best number of clusters. The distance metric d can be made to be sensitive to
the covariance structure $[\Sigma]$ of the data by defining

$$d^2(\vec{a}, \vec{b}) = ([U]\vec{a} - [U]\vec{b})^T [\Lambda]^{-1} ([U]\vec{a} - [U]\vec{b}), \qquad (5.76)$$

where $[U]$ is the matrix of eigenvectors and $[\Lambda]$ is the diagonal matrix of eigenvalues of $[\Sigma]$. The number of computations can also be reduced by discarding,
using a variety of methods, voxels that are definitely not active. Another way of
reducing the computational burden is to cluster voxel cross-correlation functions
(see Equation (1.4)) for each voxel in place of the time course [194, 196]. Extract-
ing the feature strength and delay from the cross-correlation function, as Lange
and Zeger [266] do using the gamma variate function for cross-correlation, gives
parameters that may be used in constructing more diverse feature vectors. Goutte
et al. [196] create feature vectors for each voxel by collecting parameters produced
by several activation map computation methods for a metacluster approach. They
use, in particular, a seven-dimensional feature vector composed of: (1) a simple
on–off t statistic, (2) a KS statistic, (3) a correlation statistic (with delayed stimulus
function), (4) σ of an FIR fitted signal, (5) delay from an FIR model, (6) strength
from a Lange–Zeger filter and (7) delay from a Lange–Zeger filter.

Hierarchical clustering can also begin with one cluster (the whole data set) which
is split into two using $K = 2$ in a k-means approach. Then a variety of methods,
including visual inspection of clusters in principal component space, PCA of each
cluster to assess the variability of the eigenvalues, within cluster sum of squares and
statistical similarity of the clusters using a KS statistic, may be used to determine
which cluster(s) to split in the next step [138]. These splitting criteria also provide
information on when to stop splitting the clusters.

Cordes *et al.* [104] cluster on the spectral decomposition of each voxel time
course, specifically restricting attention to frequencies less than 0.1 Hz in computing
a correlation-based distance measure to isolate activity in the resting brain.

Hierarchical clustering produces tree structures called dendrograms [330], after
the dendrites of crystal growth (see Fig. 5.5), that may themselves be manipu-
lated. Stanberry *et al.* [408] describe an approach called dendrogram sharpening.
Following Stanberry *et al.*, the dendrogram sharpening works as follows. In a

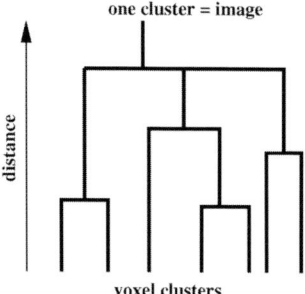

Fig. 5.5. Schematic dendrogram. At the top the minimum number of clusters is 1 which is the whole image (all the time courses). At the bottom is the maximum number of clusters in which each voxel (one time course) is a cluster. A distance between the two extremes may be defined and many different dendrograms are possible.

dendrogram, the original data points (maximal number of clusters) are called terminal nodes. Working from the terminal node up, when two clusters, the left and right child nodes, are merged they form a new node called the parent node. The ordinate value of the parent node is equal to the distance, known as an agglomeration value, between the children forming the node. The size of a node is equal to the number of terminal children under the node. At the top of the dendrogram is the root node. Stanberry *et al.* define three distance measures between clusters: (i) single linkage equal to the Euclidian distance between the nearest neighbors, (ii) complete linkage equal to the Euclidian distance between the farthest neighbors and (iii) average linkage equal to the average distance between all pairs of cluster members. The use of average-linkage leads to K-means clustering. The use of the single linkage method leads to nonoverlapping clusters. Once a dendrogram is constructed it can be sharpened in a number of ways. One is to simply delete low density clusters; this may remove small clusters. Another method is sharpening by replacement where each point is replaced by the centroid of it and its nearest neighbors. Stanberry *et al.* use an algorithm that discards all small-sized child nodes that have a large-sized parent node. Their sharpening process is controlled by two parameters, n_{fluff} and n_{core}. The parameter n_{fluff} gives the maximum size of child cluster that will be discarded if it has a parent node of size larger than n_{core}. In addition, the trimming rule also stipulates that if a particular child is subject to sharpening, it will be deleted only when its agglomeration value is larger than that of the remaining child to prevent the discarding of children composed of close children. The parameters may be used with single or multiple passes to yield different sharpened dendrograms. After dendrogram sharpening, the cluster centers need to be identified and the discarded voxels reclassified to the new cluster centers.

Baune *et al.* [37] apply a dynamical clustering algorithm (DCA) in which two parameters θ_{new} and θ_{merge} are specified in place of the number of clusters as in K-means clustering. Euclidean distance is used and the DCA method begins with an arbitrary time course vector as a cluster center, $\vec{c}_1$, and the distances to all the other data vectors are computed to that initial cluster center. The vector, $\vec{x}$, closest to the center is chosen and if $d(\vec{c}_1, \vec{x}) > \theta_{new}$, then $\vec{x}$ defines a new center $\vec{c}_2$. Otherwise, $\vec{x}$ is assigned to the first cluster and the mean of the vectors in the cluster is computed as the new cluster center. The process iterates with the additional step at each iteration that the distances between the newly computed cluster centers are computed and if any distance is less than θ_{merge}, the two corresponding clusters are merged. The parameters θ_{new} and θ_{merge} are taken as 90% and 80%, respectively, of the mode of a histogram of all pairwise distances between the data vectors.

Chen *et al.* [86] demonstrate a more sophisticated approach, that they call clustered components analysis, in which the directions of the clusters are primarily considered by clustering normalized vectors. Prior to clustered components analysis[†], Chen *et al.* reduce the dimensionality of the time courses, by projecting them into the subspace spanned by sines and cosines at the stimulus frequency (a blocked design is assumed) and its higher harmonics, further remove noise components in that harmonic subspace and finally prewhiten the data (see Section 4.1.7). For clustered components analysis, the data, $\vec{y}_n$, of voxel n are modeled as

$$\vec{y}_n = \alpha_n \vec{e}_{x_n} + \omega_n, \tag{5.77}$$

where $\vec{e}_{x_n} \in E_K = \{\vec{e}_1, \ldots, \vec{e}_K\}$ is the center of cluster x_n, $1 \leq x_n \leq K$, K is the number of clusters and ω_n is Gaussian white noise. Chen *et al.* then specify a pdf for cluster membership under the model of Equation (5.77) along with some prior probabilities $\Pi_K = \{\pi_1, \ldots, \pi_K\}$ for each cluster. Forming a Bayesean model, an EM algorithm is then used to compute the class centers and the prior cluster membership probabilities. The number of classes, K, is determined by minimizing a minimum descriptive length (MDL) criterion given by

$$\text{MDL}(K, E_K, \Pi_K) = -\ln p(y|K, E_K, \Pi_K, \hat{\alpha}) + \tfrac{1}{2} K_M \ln(N_M), \tag{5.78}$$

where K_M represents the number of scalar parameters encoded by E_K and Π_K, and N_M represents the number of scalar parameters required to represent the data.

Salli *et al.* [383, 384] describe a contextual clustering approach, based on MRF considerations, that works as follows. First a conventional (univariate) SPM$\{z\}$ map is thresholded at a threshold T_{cc} to initially divide the voxels into active and nonactive with the condition at voxel i

$$z_i > T_{cc} \tag{5.79}$$

[†] The clustered component analysis software is available from http://www.ece.purdue.edu/∼bouman.

defining an active voxel. Next the voxels are reclassified with voxel i considered active if

$$z_i + \frac{\beta}{T_{cc}}(u_i - N_n/2) > T_{cc}, \qquad (5.80)$$

where N_n is the number of voxels in a predefined neighborhood of voxel i (e.g. the 26 nearest neighbors in a 3D neighborhood), u_i is the number of currently active voxels in the predefined neighborhood and β is a parameter that weights neighborhood information and is given by

$$\beta = \frac{T_{cc}^2}{s} \qquad (5.81)$$

where s is a user selected parameter. Equation (5.80) is iterated until convergence. As $s \to 0$ contextual clustering approaches a recursive majority-vote classification, as $s \to \infty$ the method approaches voxel-by-voxel thresholding. The contextual clustering algorithm is capable of detecting activation regions whose median value is greater than T_{cc} in addition to detecting individual voxels with very high z_i. Salli *et al.* used repeated simulations to identify optimal T_{cc} and s parameters from a repeatability point of view (see Section 4.7).

Gibbons *et al.* [181] use a two level mixed effects (see Section 4.4.2) Bayesian analysis of blocked designs in which the epoch number is the second level over the voxel level. They use a third degree polynomial to fit the average HRF (as opposed to the IRF) over the epochs along with empirical Bayes, assuming multivariate normality of the polynomial coefficients, and using the EM method. Each voxel's third degree polynomial fit has two critical points (usually maximum and minimum), CP1 and CP2, which may be used to label a voxel as potentially active if the following criteria are met

1. $\text{CP1} < (t_n - t_1)/2$ and $\text{CP1} > (t_n - t_1)/4$ where t_n and t_1 are the first and last time-series points respectively. This criterion ensures that the time of maximum is in the first quartile of the epoch time interval.
2. $\text{CP2} < t_n$ and $\text{CP2} > (t_n - t_1)/(3/4)$. This criterion ensures that the time of minimum is in the third quartile of the epoch time interval.
3. $\text{CP1} - \text{CP2} \geq 2\sigma_\epsilon$ where σ_ϵ is the pooled estimate of the error variance over all voxels. This criterion gives a minimum signal to noise ratio.
4. $\text{CP1} > \bar{x} - s$ where $\bar{x}$ is the global average voxel intensity and s the corresponding standard deviation. This criterion can eliminate lower intensity, motion caused, artifact at the edge of the brain.

The resulting collection of polynomial coefficients in voxels that pass the above criteria are then subject to clustering using K-means, Gaussian finite mixture models and K-Mediod clustering methods. Gibbons *et al.* find that the nonparametric K-Mediod method provides the most homogeneous clusters.

Zang *et al.* [468] use clustering properties, rather than clustering *per se*, to create their regional homogeneity (ReHo) maps. Specifically Kendall's coefficient of concordance (see Equation (5.103)) for a pixel with its nearest neighborhood voxels is computed as

$$W = \frac{\sum_{\text{nhbd}} (R_i)^2 - n(\overline{R})^2}{K^2(n^3 - n)/12},$$ (5.82)

where K is the neighborhood size (e.g. $K = 7$, 19 or 27), R_i is the rank of time point i of n time points, and $\overline{R} = (n + 1)K/2$. The result is a W map of which one is made for each of two conditions. A t-test difference of the two W maps then shows where the activations differ between the two conditions.

Beyond clustering based on numerical information contained in the data, clustering on stimulus types is possible. Multivariate statistical pattern recognition methods, including linear discriminant analysis (a MANOVA application) and support vector machines may be applied to the response of the brain (in selected ROIs) to a given set of stimuli. Cox and Savoy [106] measure the fMRI response, in a training set of subjects, to a set of ten object pictures and classify the responses. The viewing of the objects is done in 20 s periods, not necessarily alternating, to obtain a 4D fMRI data set corresponding to the object viewed. The time course responses in specific ROIs (in the visual regions) are then classified, essentially forming a cluster of responses for each object. Then new subjects are introduced and, without knowledge of the object viewed, the response is classified to one of the previously defined clusters in an attempt to "read the brain" and determine what object was viewed.

5.4.1 Temporal cluster analysis (TCA)

TCA may be used to define a "stimulus" function, u, from the data that can subsequently be used in a standard GLM approach to compute an activation map corresponding to the defined stimulus. These TCA approaches are of use in studying the BOLD response to physiologic challenges to glucose metabolism, drug action, the onset of rapid eye movement (REM) sleep [473] or to localize epileptic activity [327]. Following Zhao *et al.* [473] TCA works by converting the (transpose of the) data matrix $[Y]$ for a v voxel data set having p time points given by

$$[Y] = \begin{bmatrix} y_{1,1} & y_{1,2} & \cdots & y_{1,p} \\ y_{2,1} & y_{2,2} & \cdots & y_{2,p} \\ \vdots & \vdots & & \vdots \\ y_{v,1} & y_{v,2} & \cdots & y_{v,p} \end{bmatrix}$$ (5.83)

into a matrix $[W]$ according to

$$w_{i,j} = \begin{cases} y_{i,j} & \text{if } y_{i,j} = \max(y_{i,1}, y_{i,2}, \ldots, y_{i,p}), \\ 0 & \text{otherwise.} \end{cases} \tag{5.84}$$

The vector

$$\vec{u} = [u_1, u_2, \ldots, u_p]^T, \tag{5.85}$$

where

$$u_j = \sum_{i=1}^{v} w_{i,j}, \tag{5.86}$$

then defines the stimulus function. Thresholding $\vec{u}$ (e.g. setting $u_j = 0$ if the total number of voxels hitting their maximum is less than a given number at time point j) gives a sharper stimulus function. TCA can also be performed with $[Y]$ replaced by a difference data set, e.g. $[Y] = [Y_1] - [Y_0]$ for two fMRI data sets obtained under conditions 0 and 1 [464]. The TCA algorithm may be run a second (and third or more) time by replacing the maximum voxel values found in the first application with zeros and applying TCA again to find the secondary maximums [176].

5.4.2 Fuzzy clustering

In fuzzy clustering the voxel time-series are given weights that define the degree of membership that the time-series has to each cluster group. This is an alternative to hard or crisp clustering in which each voxel (time-series) is assigned to a unique cluster. Fuzzy clustering has been shown to be capable of separating changes due to functional activation and other sources [443] and, when combined with appropriate model calculations, allows the quantification of flow and BOLD contributions in regions with different vascularization [328]. Fuzzy clustering has also been applied to multiple echo fMRI data sets to extract T_E dependent changes [29]. The identification of active time courses through the standard method of computing their correlation with a given model time course can result in a collection of time courses that are not necessarily correlated with each other. Fuzzy clustering avoids this problem by producing clusters of time courses that are tightly correlated with each other [31, 34]. Fuzzy clustering has also been demonstrated to outperform PCA in source separation if the data contain sources other than activation and scanner noise [33].

Let $[X] = [\vec{x}_1, \ldots, \vec{x}_n]$ be the fMRI data matrix[†], where $\vec{y}_i \in \mathbb{R}^p$ is the measured time-series for voxel i. Let $[V] = [\vec{v}_1, \ldots, \vec{v}_c]$ be the matrix of class center vectors $\vec{v}_i \in \mathbb{R}^p$ where the number of clusters, c, is fixed. Finally, let $[U]$ be the $c \times n$

[†] $[X] = [M]$ of Section 5.1 here to more closely follow the original literature.

membership matrix in which the element u_{ik} is the degree of membership (weight) of voxel i to cluster k. If d is a metric on $\mathbb{R}^p$, then $[U]$ and $[V]$ are computed by minimizing

$$J_m([X],[U],[V]) = \sum_{i=1}^{c}\sum_{k=1}^{n} u_{ik}^m d^2(\vec{x}_k,\vec{v}_i) \tag{5.87}$$

subject to the constraints

$$0 \le u_{ik} \le 1, \tag{5.88}$$

$$0 < \sum_{k=1}^{n} u_{ik} \le n, \tag{5.89}$$

$$\sum_{i=1}^{c} u_{ik} = 1, \tag{5.90}$$

where $m > 1$ is a fuzziness index used to "tune out" noise in the data [134]. Incorporating the constraints into Equation (5.87) leads to an iterative scheme for computing $[U]$ and $[V]$. Using the subscript l to denote the lth component of the vectors $\vec{y}_k$ and $\vec{v}_i$, the matrices are updated at each iterative step according to

$$v_{il} = \frac{\sum_{k=1}^{n} u_{ik}^m y_{kl}}{\sum_{k=1}^{n} u_{ik}^m}, \tag{5.91}$$

$$u_{ik} = \frac{1}{\sum_{j=1}^{c} (d(\vec{x}_k,\vec{v}_i)/d(\vec{x}_k,\vec{v}_j))^{2/m-1}}, \tag{5.92}$$

where the new quantities are on the left and the old ones are on the right. Iteration proceeds until $\max |u_{ik}(t) - u_{ik}(t-1)| < \epsilon$, where t indexes the iterations and ϵ is a convergence tolerance. The iterations are started with a random initialization matrix given by

$$[U](0) = \left(1 - \frac{1}{\sqrt{2}}\right)[U_u] + \frac{1}{\sqrt{2}}[U_r], \tag{5.93}$$

where $[U_u]$ is a matrix with every entry set to $1/c$ and $[U_r]$ is a random hard cluster assignment matrix.

Some choices for the distance measure include the Euclidean measure [30]

$$d^2(\vec{x}_k,\vec{v}_j) = \sum_{l=1}^{p}(x_{kl} - v_{jl})^2, \tag{5.94}$$

a hyperbolic correlation measure [134]

$$d(\vec{x}_k,\vec{v}_j) = [(1 - r_{jk})/(1 + r_{jk})]^\beta, \tag{5.95}$$

where r_{jk} is the correlation between $\vec{y}_k$ and $\vec{v}_j$ and where β is a fuzziness index usually set to 1 to allow m to control the fuzziness, and [188]

$$d(\vec{x}_k, \vec{v}_j) = 2(1 - r_{jk}). \tag{5.96}$$

The distance given in Equation (5.95) is a pseudometric, meaning that it cannot separate all vectors in $\mathbb{R}^p$ (the triangle inequality is not satisfied). The metric of Equation (5.96) is, however, topologically equivalent to the Euclidean metric (as are all true metrics in $\mathbb{R}^p$) and will lead to a convergent cluster solution. The convergence of clustering schemes based on pseudo-metrics needs to be checked.

To perform fuzzy clustering, it is necessary to choose valid values for the fuzziness index m and number of clusters c. The problem of choosing the correct m and c is known as the cluster validity problem and does not yet have a complete solution, although there are a number of ad hoc solutions in use. The validity of a cluster analysis may be measured by one of a number of validity functionals that give the validity of a cluster solution as a function of m and c. Frequently, m is fixed and c is varied between 2 and $c_{max} = \sqrt{n}$ or $n/3$ [134]. One such validity functional is the SCF functional given by Fadili *et al.* [135], where $SCF = SCF_1 + SCF_2$ is composed of two functionals SCF_1 and SCF_2 that quantify the clustering solution as follows. The SCF_1 measure is defined as the ratio of the compactness and average separation between the clusters and is given by

$$SCF_1 = \frac{\sum_{i=1}^c \sum_{k=1}^n \left(u_{ik}^m d^2(\vec{x}_k, \vec{v}_i) / (\sum_{k=1}^n u_{ik}) \right)}{\frac{1}{c} \sum_{i=1}^c d^2(\vec{\bar{x}}, \vec{v}_i)}, \tag{5.97}$$

where $\sum_{k=1}^n u_{ik}$ is the number of voxels in cluster i. The SCF_2 criterion expresses the fuzzy relationships between the clusters in terms of fuzzy union (FU) and fuzzy intersection (FI) as follows:

$$SCF_2 = \frac{FI}{FU} = \frac{\sum_{i=1}^{c-1} \sum_{j=i+1}^c FI_{ij}}{FU}, \tag{5.98}$$

where

$$FI_{ij} = \frac{\sum_{k=1}^n \left[\min_{i,j}(u_{ik}) \right]^2}{\sum_{k=1}^n \min_{i,j}(u_{ik})} \tag{5.99}$$

and

$$FU = \frac{\sum_{k=1}^n \left[\max_{i \in \{1,\dots,c\}}(u_{ik}) \right]^2}{\sum_{k=1}^n \max_{i \in \{1,\dots,c\}}(u_{ik})}. \tag{5.100}$$

The clustering is optimal when SCF is minimized. Resampling as a means of validating an individual cluster is also an option [35]. To apply resampling, for each vector in the cluster compute the correlations of the vector with a large number of component permutations ($\sim$10 000) of the cluster center vector, compute a p value for the vector based on the resulting empirical distribution and eliminate the corresponding voxel from the cluster if p is too high. Clusters may also be trimmed using Kendall's coefficient of concordance [32]. The trimming method works as follows. Modify the data matrix $[Y_k]$, which contains the voxel time courses of the kth cluster as columns (of which there are N) to the matrix $[Z]$ using

$$z_{ij} = (y_{ij} - \bar{y}_i)^2, \tag{5.101}$$

where $\bar{y}_i = \sum_{j=1}^{N} y_{ij}/n$ is the global cluster mean of scan i. For each voxel compute

$$V_j = \sum_{i=1}^{p} z_{ij}, \tag{5.102}$$

then it can be shown that

$$\sum_{j=1}^{N} V_j = K(1 - W), \quad K = \frac{N^2 p(p^2 - 1)}{12(N-1)}, \tag{5.103}$$

where W is Kendall's coefficient of concordance and $(1 - W)$ is the discordance. So V_j defines the contribution of $\vec{y}_j$ to the discordance of the cluster. Therefore removing the vector with the highest V_j will lower the discordance of the cluster. Vectors can be removed until a desired level of concordance is reached. That level of discordance may be expressed by stopping the trimming process when the minimum Spearmann rank correlation coefficient $r_{k,l}$ between any pair of vectors in the cluster is significant. The Spearmann rank correlation coefficient is used because it can be shown that the average Spearmann rank correlation coefficient, r_{avg} of the cluster is related to Kendall's coefficient of discordance by

$$r_{\mathrm{avg}} = \frac{NW - 1}{N - 1}. \tag{5.104}$$

A visual assessment[†] of the concordance may be obtained by stacking the data vectors as rows on top of each other and displaying as a gray scale image [36].

Another consideration in fuzzy clustering is that the constrained functional J_m of Equation (5.87) has many local minimums so that alternative methods may be needed to ensure that a global minimum is found [322]. Also, an fMRI data set contains many noisy time courses that need to be removed to make fuzzy clustering a more balanced process [134]. To reduce the size of the data set, voxels can be removed on the basis of anatomy (e.g. white matter or CSF) or with a spectral peak statistic that may be used to remove time courses with flat spectral densities (white

---

[†] For a fee, the EvIdent fuzzy clustering software, which allows the visual assessment of concordance, is available from http://www.ibd.nrc-cnrc.gc.ca/english/info_e_evident.htm.

noise voxels) [232]. Finally, after fuzzy clustering, the cluster centers can be tested using an ordinary GLM (usually a simple correlation with a model vector suffices) to determine the relationship of the cluster to the original experimental paradigm.

5.4.3 Vector quantization (VQ) and neural-network-based clustering

VQ is a general signal processing approach that is applied to a data set of vectors (signals – in fMRI, the data set of voxel time-series). VQ clustering identifies several groups, or clusters, in the data set consisting of similar, by some measure, vectors. The groups or clusters are represented by prototypical vectors called codebook vectors (CV) that represent the center of the clusters. Each data vector may be assigned to a definite cluster in a "crisp" assignment scheme or may be given weights for each cluster in a "fuzzy" assignment scheme. VQ approaches determine cluster centers, $\vec{w}_i$ by making them analogous with neural net connection weights that are iteratively updated, at times t, by a learning rule, for the ith neuron or cluster center, of the form [446]

$$\vec{w}_i(t+1) = \vec{w}_i(t) + \epsilon(t) \, a_i(\vec{x}(t), C(t), \kappa) \, (\vec{x}(t) - \vec{w}_i(t)), \qquad (5.105)$$

where $\vec{\epsilon}(t)$ represents the learning parameter, $\vec{x}$ is a randomly chosen time course and a_i is a function dependent on a codebook dependent cooperativity function $C(t)$ and a cooperativity parameter κ. Examples of VQ methods include [446]

(i) *Kohonen's self-organizing map (SOM)* [92, 139, 339] in which the neurons are arranged on a uniform square or hexagonal lattice with fixed distances d_{ij} between the nodes[†]. Time course centers are organized in a meaningfully and visually obvious way on the lattice after classification, see Fig. 5.6. The cooperativity function for the learning rule for Kohonen's SOM is

$$a_i = \exp(-d_{ij}/\sigma^2), \qquad (5.106)$$

where σ^2 is an operating parameter. Chuang *et al.* [92] follow up a Kohonen SOM with a fuzzy clustering analysis of the SOM components to reduce redundancy.

(ii) *Fuzzy clustering based on deterministic annealing.* The cooperativity function is the softmax activation function given by

$$a_i = \frac{\exp(-\|\vec{x}(t) - \vec{w}_i(t)\|^2/2\rho^2)}{\sum_i \exp(-\|\vec{x}(t) \quad \vec{w}_i(t)\|^2/2\rho^2)}, \qquad (5.107)$$

where ρ is a fuzziness parameter and $2\rho^2$ may be interpreted as a temperature. The algorithm starts with one cluster (the whole data set) and splits them as the iterations proceed.

[†] The public domain software SOM PAK is available at http://citeseer.ist.psu.edu/kohonen96som.html.

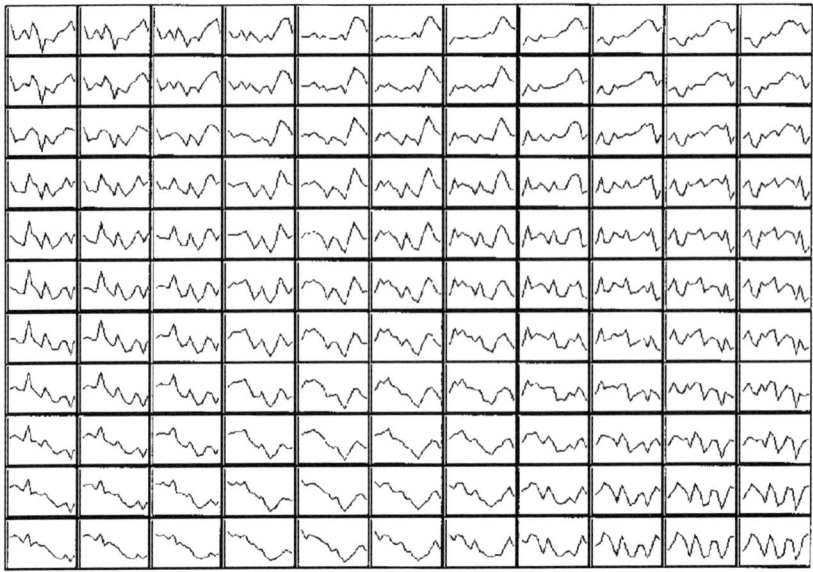

Fig. 5.6. Example time courses extracted by a Kohonen SOM approach. The fMRI experiment involved periodic visual stimulation. Notice how the time courses are related to each other both along the rows and along the columns. Taken from [139], used with permission.

(iii) *"Neural gas" network.* The cooperativity function is

$$a_i = \exp(-k_i(\vec{x}, \vec{w}_i)/\lambda), \qquad (5.108)$$

where $k_i = 0, 1, \ldots, N - 1$ represents a rank index of the reference vectors $\vec{w}_i$ to $\vec{x}$ in a decreasing order, N is the number of neurons in the network and λ determines how many neurons will change their synapses in an iteration.

Wismüller *et al.* [446] discuss the relative merits of the three above VQ methods.

5.4.4 Replicator dynamics

Lohmann and Bohn [286] introduce the method of replicator dynamics to fMRI for the purpose of finding functionally connected clusters or networks of voxels in the brain. Replicator dynamics begins with a $v \times v$ similarity matrix $[W]$, where v is the number of voxels in the data set. The similarity matrix may be given by correlations (computed, for example, from $[M]^T[M]$, where $[M]$ is the fMRI data matrix) or by a matrix of Spearman's rank correlation coefficients between the voxel time courses or by mutual information. The similarity matrix needs to contain positive entries in order for the replicator dynamics algorithm to converge, so correlation-based $[W]$ need to either use the absolute values of the correlation coefficients or set negative correlations to zero. A network, a set of voxels N, found by replicator dynamics is defined by a vector $\vec{x}$ in a unit v-dimensional hypercube. Each component of $\vec{x}$ represents a voxel and a value of 0 means that the voxel is not a part of N and a value of 1 indicates membership in N. Values in $(0, 1)$ indicate fuzzy membership in N. A network is defined as the solution to the following dynamical system

$$\frac{d}{dt} x_i(t) = x_i(t)[([W]\vec{x}(t))_i - \vec{x}(t)^T[W]\vec{x}(t)], \quad i = 1, \ldots, v, \tag{5.109}$$

where the subscript i denotes the ith component. The discrete (in time) version of Equation (5.109) is given by

$$x_i(t+1) = x_i(t)\frac{([W]\vec{x}(t))_i}{\vec{x}(t)^T[W]\vec{x}(t)} \tag{5.110}$$

and Equations (5.109) and (5.110) are known as replicator equations. A fundamental theorem of natural selection guarantees that replicator systems will converge to a stationary state, our desired network N. The process of evolving the replicator dynamics is begun from the state in which $x_i = 1/v$ for all i. After convergence, the network is "defuzzified" by setting all $x_i > 1/n$ equal to 1 and the rest to 0. After one network is found, another network may be found by considering only those voxels that are not contained in the first network. The process is repeated until a desired number of networks is found. After the networks are found, it is possible to update $[W]$ by setting the similarity value, an entry in $[W]$, of voxel i equal to the average of all the similarity values of the voxels that belong to the same network as voxel i. Then the networks may be recomputed using the updated $[W]$. To use the replicator dynamics approach for a group of n subjects, an average $[W]$ may be employed where the average is computed using Fisher's z transform. That is, each entry, r, in the similarity matrix of an individual subject is transformed to $z = f(r)$, where

$$z = f(r) = 0.5 \ln \left(\frac{1+r}{1-r} \right). \tag{5.111}$$

The inverse transformation is given by

$$r = f^{-1}(z) = \frac{e^{2z} - 1}{e^{2z} + 1}, \tag{5.112}$$

so if $[W_i]$ is the similarity matrix for subject i, then the group similarity matrix $[W]$ is given, entry-wise, by

$$[W] = f^{-1}\left[\sum_{i=1}^{n} f([W_i]) \right]. \tag{5.113}$$

To visualize the similarity structure given by $[W]$, Lohmann and Bohn use multi-dimensional scaling (MDS) [159]. For time courses comprising k time points, MDS

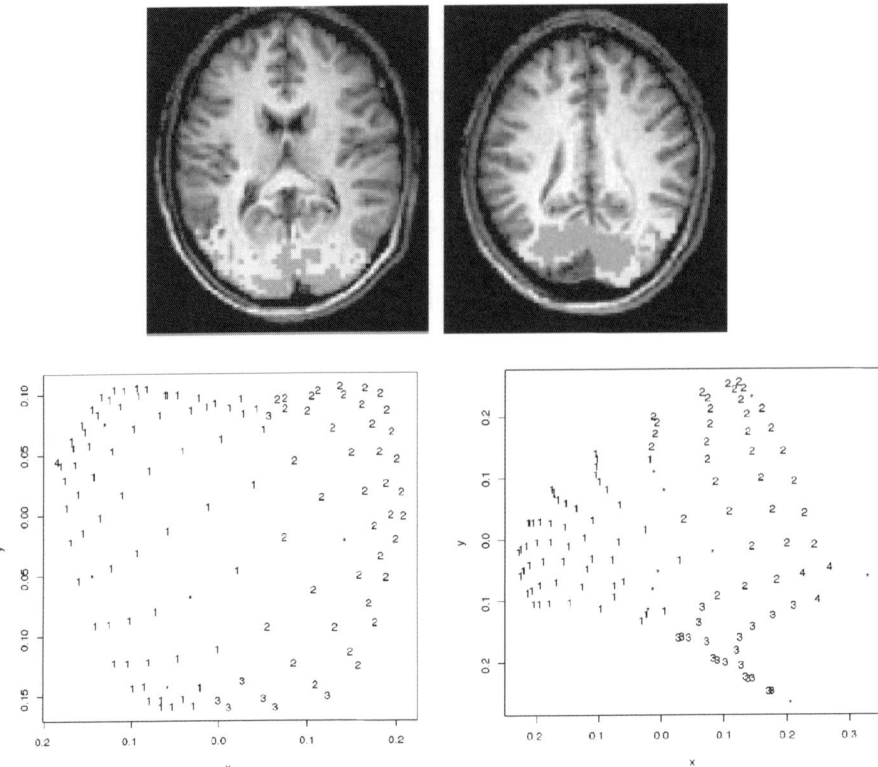

Fig. 5.7. Example replicator dynamics solution to finding networks in a visual stimulation experiment. At the top, two slices from one subject are shown with the found networks color coded with red representing the first network and various shades of yellow representing other networks. At the bottom are the MDS maps (plotted with the two principal directions of the similarity structure matrix [W] represented as the x and y directions) for each slice with the networks being represented by their numbers. In this example, SOM clustering was done first, followed by the determination of networks of cluster centers by replicator dynamics from the SOM clusters. Note how the SOM organization shows up in the MDS clusters as expected. Taken from [286], ©2002 IEEE, used with permission. See also color plate.

is a way of visualizing clusters of time courses in the unit cube in $\mathbb{R}^k$. With the similarity measures in [W] representing distances in $\mathbb{R}^k$, normalized time courses are plotted in $\mathbb{R}^k$, usually with the axes along the principal directions of [W] so that structure may be maximally seen, see Fig. 5.7. The network clusters found should be close in distance in $\mathbb{R}^k$.

A variation for finding the second network is to use the eigendecomposition of [W] given by

$$[W] = [P]^T [\Lambda][P], \tag{5.114}$$

where $[P]$ is a matrix of the eigenvectors of $[W]$ and $[\Lambda]$ is a diagonal matrix containing the eigenvalues. The first eigenvalue of $[\Lambda]$ is set to 0 to give $[\Lambda_2]$. Then the similarity matrix given by

$$[W_2] = [P]^T [\Lambda_2][P] \tag{5.115}$$

is used on the whole data set in the replicator equation to find the second network. In this way, the second network may contain voxels from the first network. Finally, Lohmann and Bohn show that replicator dynamics may be used to find networks of cluster centers identified in SOM clustering.

5.5 Functional connectivity

The component decomposition and clustering methods reviewed so far characterize functional connectivity primarily in terms of correlations between voxel time courses. For example, ICA uses the information theoretic concept of mutual information which may be related to correlation between voxels in the following way [147]. Let p and q represent two ROIs of voxels represented by data matrices $[M_p]$ and $[M_q]$. Then the mutual information between the two regions, MI_{pq}, is given by

$$MI_{pq} = \ln \left(\frac{|[M_p]^T[M_p]| \ |[M_q]^T[M_q]|}{\sqrt{|[M]^T[M]|}} \right), \tag{5.116}$$

where $| \cdot |$ denotes a determinant. In the limiting case in which each ROI consists of one voxel

$$MI_{pq} = - \ln(1 - r^2)/2, \tag{5.117}$$

where r is the correlation between the two voxel time-series.

A few groups have developed and applied a method they call functional connectivity MRI (fcMRI) based on the measurement of low frequency (< 0.1 Hz) signal in the resting brain [45, 103, 197, 205, 288, 462]. This resting brain fcMRI works by first having the subject perform a task that will activate a region of interest such as a part of the motor cortex (e.g. from finger tapping) or a speech region like Broca's area. Then an fMRI time-series is obtained with the subject resting through the whole series. A seed voxel time course is then selected from the active ROI(s) identified in the task activation analysis and all voxels correlated with that seed time course are declared to be a part of the resting network associated with the original ROI. Care must be taken that fcMRI is not measuring aliased cardiac or respiratory signal [291] or that some other phenomenon besides neural activity induced BOLD is being observed [295]. However, it is worth noting that measurements of flow made using the spin tagging flow-sensitive alternating inversion recovery (FAIR) method have shown that fcMRI resting state signals are BOLD-based and not blood-flow-based [48]. The null distributions of the correlation coefficients found in fcMRI may be determined by permutation in the wavelet domain (where the time-series noise is uncorrelated) so that significantly correlated voxels may be identified [56].

Clusters of functionally connected regions (general functional connectivity now, not just resting brain studies) may be assessed for statistical significance using random field theory for cross-correlation and autocorrelation fields similarly to how Gaussian random fields are used to assess the significance of clusters with univariate SPM [456] (see Section 4.1.8). Given two fMRI time-series data sets containing n time points each, with the individual voxel values denoted by $X_i(x_1, y_1, z_1)$ and $Y_i(x_2, y_2, z_2)$, then their 6D cross-correlation field, R, is given by

$$R(x_1, y_1, z_1, x_2, y_2, z_2) = \frac{\sum_{i=1}^{n} X_i(x_1, y_1, z_1) Y_i(x_2, y_2, z_2)}{\sqrt{\sum_{i=1}^{n} X_i(x_1, y_1, z_1)^2} \sqrt{\sum_{i=1}^{n} Y_i(x_2, y_2, z_2)^2}}. \quad (5.118)$$

An autocorrelation field is defined with $X = Y$. Such random field methods may be used in lieu of SVD, CVA, PLS or MLM (see Section 5.2.4) to assess correlation-based functional connectivity.

Functional networks or clusters are identified by d'Avossa *et al.* [23] on the basis of BOLD response shapes from a blocked experiment. Their procedure is as follows. First, active regions are identified using univariate SPM methods. Then a PCA analysis is computed and the nonnoisy components (as determined from a separate SNR estimate) are kept. The time courses of the active voxels are projected into the space spanned by the retained principal components to reduce data dimensionality. The dimension-reduced, active time courses are then subjected to a fuzzy cluster analysis with the number of clusters being set by a cluster validation procedure based on mutual information. The fuzzy cluster identified time courses serve as the prototypical BOLD time course and the clusters represent functionally connected regions.

Sun *et al.* [414] use coherence between time-series as a Fourier-based measure of correlation for finding functionally connected regions. Specifically, if f_{xx}, f_{yy} and f_{xy} represent the power spectra of time-series x and y and the cross-spectrum of x and y respectively, then the coherency $\text{Coh}_{xy}(\lambda)$ of x and y at frequency λ is given by

$$\text{Coh}_{xy}(\lambda) = |R_{xy}(\lambda)|^2 = \frac{|f_{xy}(\lambda)|^2}{f_{xx}(\lambda) f_{yy}(\lambda)}, \quad (5.119)$$

where R_{xy} is the complex-valued coherency of x and y. To avoid detecting correlations due to an external source and not due to interactions between the voxels (stimulus-locked response) a partial coherence, taking the external reference time course r into account, as given by

$$\mathrm{Coh}_{xy|r}(\lambda) = \frac{|R_{xy}(\lambda) - R_{xr}(\lambda)R_{ry}(\lambda)|^2}{(1 - |R_{xr}(\lambda)|^2)(1 - |R_{ry}(\lambda)|^2)} \tag{5.120}$$

may be used. Sun *et al.* use the seed voxel method with coherence to identify functionally connected clusters.

5.6 Effective connectivity

Friston [147] defines the idea of effective connectivity as being distinct from functional connectivity with the idea that effective connectivity depends on some model of the influence that one neuronal system exerts on another. With effective connectivity, we can begin to study brain dynamics as opposed to simply identifying brain regions associated with certain functions in an approach some refer to as neo-phrenology [149]. It is only with these deeper models of brain function that incorporate effective connectivity that we can hope to advance the understanding of mysteries like memory and consciousness using fMRI and neuroimaging in general [390]. The development of models and understanding of connectivity (both functional and effective) depends on the use of information from other imaging sources (e.g. PET, MEG, EEG) and, as such, a thorough understanding of what connectivity means for each modality and of the analysis procedure is needed for the interpretation of experimental results [224, 273]. With the MRI modality, diffusion tensor imaging (DTI) has been effectively used to construct and verify connectivity models on the basis of the resulting white-matter fiber tract maps provided by DTI [256, 398]. The validation of models of effective connectivity is important and Friston [147] identifies three types of validity concerns. These are construct validity (does the model have validity in terms of another construct or framework), face validity (does the model capture what it is supposed to) and predictive validity (does the model accurately predict the system's behavior).

Let us begin our review of effective connectivity models with the simple model originally proposed by Friston [147]. Let $[M]$ be the $t \times v$ fMRI data matrix for t scans and v voxels and let $\vec{m}_i$ be the time course of voxel i. Then the simple model expresses the time course at voxel i as a linear combination of the time courses of all the other voxels:

$$\vec{m}_i = [M]\vec{C}_i + \vec{\epsilon}_i, \tag{5.121}$$

where $\vec{C}_i$ is the vector of effective connectivities from all the voxels to voxel i and $\vec{\epsilon}_i$ is an error vector. The least squares solution to Equation (5.121) is given by

$$\vec{C}_i = ([M]^T[M])^{-1}[M]^T\vec{m}_i. \tag{5.122}$$

In fact, defining the effective connectivity matrix $[C]$ as the matrix whose columns are the vectors $\vec{C}_i$, the least squares solution for the effective connectivity matrix is

$$[C] = [V][V]^T,   \quad (5.123)$$

where $[V]$ is as given by Equation (5.1). In contrast, the functional connectivity matrix is given by

$$[M]^T[M] = [V][S]^2[V]^T.   \quad (5.124)$$

So effective connectivity, in this model, is represented by orthogonal modes (eigenimages) with internal unit connections between the voxels in a mode. With functional connectivity, the modes are weighted by the variance accounted for (eigenvalues) by each mode. One immediate face validity issue with the simple model is that $[C]$ is symmetric. That is, the connections between the voxels are two-way with the strength in each direction being equal. Nonlinear models with one-way or unequal strength back connections may be modeled by structural equation modeling (SEM) (see Section 5.6.2).

5.6.1 Mental chronometry revisited

One obvious way of modeling effective connectivity is to order activated regions on the basis of their latency under the assumption that one region must be activated before another can be activated by it [318]. In this respect, the methods reviewed in Section 4.2.3 are useful. Here we review a couple of multivariate approaches that incorporate BOLD timing.

Working in the spectral domain, Müller *et al.* [331] use coherence as defined in Equation (5.119) in addition to the phase lead $\theta_{jk}(\lambda)$ of voxel j over k as defined by

$$f_{jk}(\lambda) = |f_{jk}(\lambda)| \, e^{i\theta_{jk}(\lambda)}   \quad (5.125)$$

at the frequency λ of the blocked presentation design. The phase lead at the presentation frequency is readily converted to a latency (or time to peak) difference that may be used to order functionally connected (coherent) regions temporally.

Lahaye *et al.* [262] define a functional connectivity measure that includes, in addition to correlations between voxel time courses, the influence of the past history of one time course on another. The functional connectivity maps so defined therefore include more effective connectivity information than functional connectivity maps based on correlation (which reveals linear instantaneous interactions only) alone. Lahaye *et al.* produce a sequence of nested connectivity models that range from the standard correlation model to one that includes both hemodynamic delay (history) effects and nonlinear effects. Their models are applied to a voxel data set that is reduced in the number of voxels by a parcellation method that produces 100 equal volume parcels comprising gray matter only. Thus only 100 parcel time courses, obtained preferentially by averaging the original time courses in the parcel, are analyzed. Let $x_a(t)$ and $x_b(t)$ be the time courses of parcels a and b respectively and let $\mathcal{G}^i_{b \to a}$ denote the model regressor for the directed interaction $b \to a$ which is described by the model

$$x_a = \mathcal{G}^i_{b \to a} \cdot \beta^i_{b \to a}. \tag{5.126}$$

Based on their finding that the history, in addition to correlation, is important for defining functional connectivity, Lahaye *et al.* recommend that functional connectivity be based on an F statistic obtained by comparing an AR model, $\mathcal{M}^{\mathrm{AR}}_{b \to a}$ and a history model, $\mathcal{M}^{\mathrm{HD}}_{b \to a}$ defined by the regressors

$$\mathcal{G}^{\mathrm{AR}}_{b \to a} = \{x_a(t - k\tau)_{1 < k < M}\}, \tag{5.127}$$

$$\mathcal{G}^{\mathrm{HD}}_{b \to a} = \{x_a(t - k\tau)_{1 < k < M}, x_b(t), x_b(t - k\tau)_{1 < k < M}\}, \tag{5.128}$$

where $\tau = T_R$. The inclusion of $x_b(t)$ models correlation, while $x_b(t - k\tau)_{1 < k < M}$ models prior history effects. Lahaye *et al.* note that their models respect Granger causality, while Volterra-kernel-based models, for nonlinear modeling, may not. Granger causality states that in the $b \to a$ causality, x_a depends on past values of x_b and is independent of future values of x_b.

5.6.2 Structural equation modeling (SEM)

SEM, also known as path analysis, postulates causal relationships between observed or measured variables and unobserved or latent variables. Applications of SEM to date have not used latent variables, but we will give a general outline of SEM here because it is possible that latent variables could be used to model underlying neural activity [418]. Latent variables have a similar standing to factors which we have seen can be obtained from PCA (see Equation (5.22)). There are several ways to specify an SEM model; we will follow the LISREL model[†] here because of its simplicity. Define two measured variables, $\vec{x}$, the independent measured variable (IV), and $\vec{y}$, the dependent measured variable (DV). The vectors have as their components, in the fMRI case, observed time course vectors (with the mean subtracted so that the time courses represent deviations from the mean), one component per region. Similarly, define the IV and DV latent variables by $\vec{\xi}$ and $\vec{\eta}$ respectively. Then the full LISREL model consists of three submodels: (1) the structural equation model

$$\vec{\eta} = [B]\vec{\eta} + [\Gamma]\vec{\xi} + \vec{\zeta}, \tag{5.129}$$

[†] LISREL = LInear Structural RELations is implemented in a commercial software package available from http://www.ssicentral.com/lisrel/mainlis.htm. There are other commercial SEM software packages available, none of them are specifically designed for fMRI. The freely available SPM software does, however, have an SEM toolbox.

(2) the measurement model for $\vec{y}$

$$\vec{y} = [\Lambda_y]\vec{\eta} + \vec{\epsilon}, \tag{5.130}$$

and (3) the measurement model for $\vec{x}$

$$\vec{x} = [\Lambda_x]\vec{\xi} + \vec{\delta}. \tag{5.131}$$

Note that the two measurement models are essentially factor models. Further, define the covariance matrices

$$\text{Cov}(\vec{\xi}) = [\Phi] \quad \text{Cov}(\vec{\zeta}) = [\Psi] \quad \text{Cov}(\vec{\epsilon}) = [\Theta_\epsilon] \quad \text{Cov}(\vec{\delta}) = [\Theta_\delta]. \tag{5.132}$$

The entries in the (in general unsymmetrical) matrices $[B]$ and $[\Gamma]$ define the path coefficients that represent the causal influence of one variable on another. The matrix $[B]$ is restricted to have zeros on the diagonal and it is assumed that $\vec{\epsilon}$ is uncorrelated with $\vec{\eta}$, $\vec{\delta}$ is uncorrelated with $\vec{\xi}$, $\vec{\zeta}$ is uncorrelated with $\vec{\xi}$ and $\vec{\zeta}$, $\vec{\epsilon}$ and $\vec{\delta}$ are mutually uncorrelated. Under these assumptions and constraints, the covariance matrix implied by the full model is given by

$$[\Sigma] = \begin{bmatrix} [\Lambda_y][A]([\Gamma][\Phi][\Gamma]^T + [\Psi])[A]^T[\Lambda_y]^T + [\Theta_\epsilon] & [\Lambda_y][A][\Gamma][\Phi][\Lambda_x]^T \\ [\Lambda_x][\Phi][\Gamma]^T[A]^T[\Lambda_y]^T & [\Lambda_x][\Phi][\Lambda_x]^T + [\Theta_\delta] \end{bmatrix}, \tag{5.133}$$

where $[A] = ([I] - [B])^{-1}$. The goal is then to find $[B]$, $[\Gamma]$, $[\Lambda_y]$ and $[\Lambda_x]$ such that the difference between the data covariance matrix and $[\Sigma]$ of Equation (5.133) is minimized. The significance of the resulting path coefficients may then be assessed using an appropriate χ^2 statistic. The SEM model is almost always represented by a path diagram, which for the LISREL setup has IVs on the left, DVs on the right, measured variables on the outside and latent variables on the inside, see Fig. 5.8.

For application to fMRI the latent IVs and DVs are identified with the measurement IVs and DVs respectively so Equations (5.130) and (5.131) reduce to $\vec{y} = \vec{\eta}$ and $\vec{x} = \vec{\xi}$ and we are left with a single structural equation model

$$\vec{y} = [B]\vec{y} + [\Gamma]\vec{x} + \vec{\zeta}. \tag{5.134}$$

To apply SEM to fMRI an explicit model needs to be constructed based on previous information on how different brain regions interact. In this respect DTI [201, 353, 441] should be very useful for constructing SEM models, especially in patients whose connective pathways are physically absent. Most SEM approaches to date select a relatively small number of regions ($\sim$3–6) to model. The data representing each region then need to be defined. The time courses or their projection onto the first PCA component in the active region may be averaged together, or the time course or its projection onto the first PCA component from the voxel with

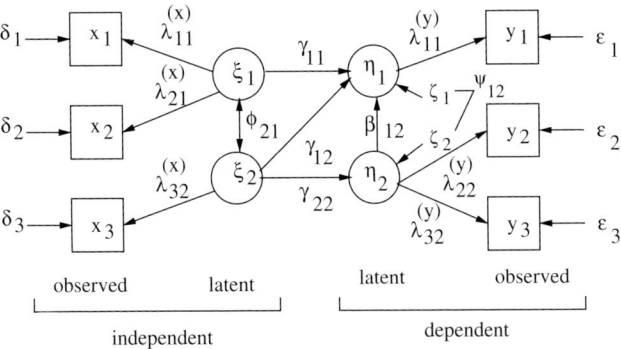

Fig. 5.8. A generic LISREL path diagram. From left to right, δ_i are components of the error $\vec{\delta}$, x_i are components of the observed independent $\vec{x}$, $\lambda_{ij}^{(x)}$ are entries in the measurement model matrix $[\Lambda_x]$, ξ_i are the latent independent variables in $\vec{\xi}$, ϕ_{ij} are entries in the (symmetric) $[\Phi]$, γ_{ij} are the path coefficients in $[\Gamma]$, β_{ij} are the path coefficients in $[B]$, η_i are the dependent latent variables in $\vec{\eta}$, ζ_i are components of the error $\vec{\zeta}$, ψ_{ij} are entries in the (symmetric) $[\Psi]$, $\lambda_{ij}^{(y)}$ are entries in the measurement model matrix $[\Lambda_y]$, y_i are the observed dependent variables in $\vec{y}$ and ϵ_i are components of the error $\vec{\epsilon}$.

the highest SPM t statistic may be used [190]. Honey *et al.* [221] further use only the time points associated with the task, discarding the rest time points. Variations in the data representation used and exact ROI definition lead to different solutions for the SEM model [190]. Nevertheless, SEM has been applied to study differences in learning between normal and schizophrenic subjects [392], differences in verbal working memory load [221] and changes due to learning [61] to name a few examples. SEM may also be used in group analysis with a virtual stimulus presentation node connected to all subjects [309]. Büchel and Friston have used a series of SEM models to show that "top-down" attention can modulate the "bottom-up" visual processing pathway [58, 62, 63], an effect they verified using direct nonlinear modeling using Volterra series [167].

We use one of Büchel and Friston's SEM models [63] to illustrate a path diagram and its relation to Equation (5.134). The path diagram is illustrated in Fig. 5.9. The corresponding SEM is given by

$$\begin{pmatrix} PP \\ V5 \\ V1 \end{pmatrix} = \begin{bmatrix} 0 & 0.69 & 0 \\ 0 & 0 & 0.87 \\ 0 & 0 & 0 \end{bmatrix} \begin{pmatrix} PP \\ V5 \\ V1 \end{pmatrix} + \begin{bmatrix} 0 & 0 \\ 0.14 & 0 \\ 0 & 0.61 \end{bmatrix} \begin{pmatrix} V1 * PP \\ LGN \end{pmatrix} + \begin{pmatrix} \zeta_1 \\ \zeta_2 \\ \zeta_3 \end{pmatrix},$$

(5.135)

where we have represented the time courses with a notation that represents the regions defined in the caption to Fig. 5.9. This particular SEM has a nonlinear

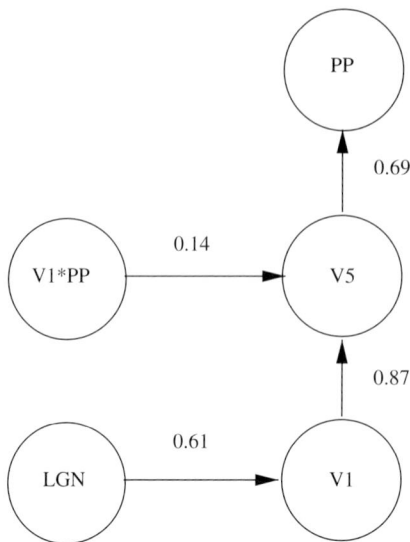

Fig. 5.9. An example path diagram for a structural equation model (SEM) presented by Büchel and Friston [63]. The independent variables, $\vec{x}$, are drawn on the left, the dependent variables, $\vec{y}$, are drawn on the right. The ROI definitions are: LGN = lateral geniculate nucleus, V1 = primary visual cortex, V5 = motion sensitive area and PP = posterior parietal complex. The term V1*PP is an interaction term. The SEM corresponding to the path diagram is given in Equation (5.135).

variable, V1*PP, constructed by multiplying the time courses from V1 and PP to represent their interaction.

5.6.3 *Vector autoregressive modeling (VAR)*

Suppose that we are interested in the effective connectivity between M ROIs. Let $\vec{x}(n) = (x_1(n), \ldots, x_M(n))^T$ denote the measured response in those M ROIs at time n. With SEM the value of any given $x_i(n)$ can depend only on linear combinations of the values of the other components at time n. VAR extends this model to make $\vec{x}(n)$ a function of $\vec{x}(t)$ for $t < n$ which allows the history of the responses to influence the current response. Explicitly, a vector time-series $\vec{x}(n)$ can be modeled as a VAR(p) process as

$$\vec{x}(n) = -\sum_{i=1}^{p} [A(i)]\vec{x}(n-i) + \vec{u}(n), \qquad (5.136)$$

where $\vec{u}(n)$ is multivariate white noise with $\text{var}(\vec{u}(n)) = E(\vec{u}(n)\,\vec{u}^T(n)) = [\Sigma]$. Goebel *et al.* [187] construct a series of three VAR models and use the covariance matrices from these models to define a measure of influence $F_{x,y}$ between an activated reference region, represented by K ROIs and $\vec{x}(n) = (x_1(n), \ldots, x_K(n))^T$ and other activated regions, represented by L ROIs and $\vec{y}(n) = (y_1(n), \ldots, y_L(n))^T$. Defining $\vec{q}^T(n) = (\vec{x}^T(n), \vec{y}^T(n))$, the three VAR models are

$$\vec{x}(n) = -\sum_{i=1}^{p}[A_x(i)]\vec{x}(n-i) + \vec{u}(n) \quad \text{var}(\vec{u}(n)) = [\Sigma_1], \tag{5.137}$$

$$\vec{y}(n) = -\sum_{i=1}^{p}[A_y(i)]\vec{y}(n-i) + \vec{v}(n) \quad \text{var}(\vec{v}(n)) = [T_1], \tag{5.138}$$

$$\vec{q}(n) = -\sum_{i=1}^{p}[A_q(i)]\vec{q}(n-i) + \vec{w}(n) \quad \text{var}(\vec{w}(n)) = [Y], \tag{5.139}$$

where

$$[Y] = \begin{bmatrix} [\Sigma_2] & [C] \\ [C]^T & [T_2] \end{bmatrix}. \tag{5.140}$$

The measure of influence $F_{x,y}$, given by

$$F_{x,y} = \ln(|[\Sigma_1]| \, |[T_1]|/|[Y]|), \tag{5.141}$$

where $|\cdot|$ denotes a determinant, may be decomposed into

$$F_{x,y} = F_{x \rightarrow y} + F_{y \rightarrow x} + F_{x \cdot y}, \tag{5.142}$$

where $F_{x \rightarrow y}$ denotes the influence of x on y, $F_{y \rightarrow x}$ denotes the influence of y on x and $F_{x \cdot y}$ denotes the instantaneous influence of x on y. The three influences are defined by

$$F_{x \rightarrow y} = \ln(|[T_1]|/|[T_2]|), \tag{5.143}$$

$$F_{y \rightarrow x} = \ln(|[\Sigma_1]|/|[\Sigma_2]|), \tag{5.144}$$

$$F_{x \cdot y} = \ln(|[\Sigma_2]| \, |[T_2]|/|[Y]|) \tag{5.145}$$

and each may be plotted, for a given reference region, to give a "Granger causality map".

Harrison *et al.* [212] take a different approach and extend the VAR model of Equation (5.136) to a nonlinear VAR by augmenting the vector $\vec{x}(n)$ with new variables (components) of the form

$$I_{j,k}(n) = x_j(n)x_k(n) \tag{5.146}$$

to form the augmented vector $\tilde{\vec{x}}(n)$ and the nonlinear VAR model given by

$$\tilde{\vec{x}}(n) = -\sum_{i=1}^{p}[\tilde{A}(i)]\tilde{\vec{x}}(n-i) + \tilde{\vec{u}}(n). \tag{5.147}$$

The relevant entries in $[\tilde{A}(i)]$ then model nonlinear interactions between regions j and k. The matrix $[\tilde{A}(i)]$ may be estimated by least squares methods (ML) and the optimal order p may be determined by Bayesian methods. VAR models may also be constrained, by regularization, to form smooth fields when used at the voxel level [428].

5.6.4 Variable parameter regression and Kalman filtering

Büchel and Friston [59] introduce a variable parameter model for characterizing the effective connectivity between two regions as a function of time. The variable parameter regression model for characterizing the influence of n regions $\vec{x}_t$ at time $1 \le t \le T$ on region y_t is given by

$$y_t = \vec{x}_t^T \vec{\beta}_t + u_t, \qquad (5.148)$$

where $u_t \sim N(0, \sigma^2)$ and $\vec{\beta}_t$ specifies the effective connectivities at time t. The time course for the influenced region is $(y_1, \ldots, y_T)$, while the time course of influencing region j is given by the jth components of the $\vec{x}_t$s as $(x_{j,1}, \ldots, x_{j,T})$. The effective connectivity time course from region j to y is given by the components $(\beta_{j,1}, \ldots, \beta_{j,T})$.

The evolution of $\vec{\beta}$ is assumed to follow a random walk with zero drift over time as given by

$$\vec{\beta}_t = \vec{\beta}_{t-1} + \vec{p}_t, \qquad (5.149)$$

where $p_t \sim N(0, \sigma^2[P])$ is an innovation (underlying stochastic process). It is assumed that u_t and $\vec{p}_t$ are uncorrelated. The parameters of the variable parameter regression model are estimated iteratively using Kalman filtering. To see how the Kalman filtering works in this case we follow Büchel and Friston and first assume that $[P]$ and σ^2 are known – they will be estimated in a second iterative loop. Let $\hat{\vec{\beta}}_t(s)$ be the estimate of $\vec{\beta}_t$ based on s observations of y and let $\sigma^2[R_t]$ be the estimated covariance matrix of $\hat{\vec{\beta}}_t(t-1)$. Then the estimate of $\hat{\vec{\beta}}_t(t)$ is obtained inductively from the estimate of $\hat{\vec{\beta}}_{t-1}(t-1)$ beginning with

$$\hat{\vec{\beta}}_t(t-1) = \hat{\vec{\beta}}_{t-1}(t-1), \qquad (5.150)$$

$$[R_t] = [S_{t-1}] - [I] + [P], \qquad (5.151)$$

where $\sigma^2[S_{t-1}]$ is the estimated covariance matrix of $\hat{\vec{\beta}}_{t-1}(t-1)$. The Kalman filter $\vec{K}_t$ revises this estimate of $\vec{\beta}_t$ with

$$\hat{\vec{\beta}}_t(t) = \hat{\vec{\beta}}_t(t-1) + \vec{K}_t e_t, \qquad (5.152)$$

where

$$e_t = y_t - \vec{x}_t^T \hat{\vec{\beta}}_t(t-1), \tag{5.153}$$

$$\vec{K}_t = [R_t]\vec{x}_t E_t^{-1}, \tag{5.154}$$

$$E_t = \vec{x}_t^T [R_t]\vec{x}_t + 1, \tag{5.155}$$

$$[S_t] = [R_t] - \vec{K}_t \vec{x}_t^T [R_t]. \tag{5.156}$$

The estimates $\hat{\vec{\beta}}_t(t)$, which are based on only t pieces of information are updated to smoothed estimates $\hat{\vec{\beta}}_t(T)$ based on information from all the data using

$$\hat{\vec{\beta}}_t(T) = \hat{\vec{\beta}}_t(t) + [G_t](\hat{\vec{\beta}}_{t+1}(T) - \hat{\vec{\beta}}_t(t)), \tag{5.157}$$

where

$$[G_t] = [S_t]([S_t] + [P])^{-1}. \tag{5.158}$$

The smoothed covariance matrix estimates are obtained from

$$[V_t] = [S_t] + [G_t]([V_{t+1}] - [R_{t+1}])[G_t]^T, \tag{5.159}$$

$$[V_T] = [S_T]. \tag{5.160}$$

The variance parameters σ^2 and $[P]$ are obtained by maximizing the log-likelihood function

$$L = -\frac{1}{2} \sum_{t=n+1}^{T} \ln(\sigma^2 E_t) - \frac{1}{2} \sum_{t=n+1}^{T} \frac{e_t^2}{\sigma^2 E_t}. \tag{5.161}$$

The log-likelihood statistic may also be used to formulate a test statistic that asymptotically has a χ^2 distribution with one degree of freedom under the null hypothesis.

5.6.5 *Dynamic causal modeling (DCM)*

DCM [173, 312] represents the most comprehensive model of brain activity based on fMRI data used to date. The basic approach is to model the neuronal-hemodynamic activity in a state space model

$$\frac{d\vec{x}}{dt} = f(\vec{x}, u, \vec{\theta}), \tag{5.162}$$

$$\vec{y} = \lambda(\vec{x}) \tag{5.163}$$

at the voxel (or small ROI) level using the model of Equations (4.158)–(4.162) with the input u to one region being either the output from another region or an exogenous influence like a stimulus or a category of stimulus. A bilinear connectivity model of the neuronal activity z (z is a component of $\vec{x}$ in Equation (5.162)) underlying a single voxel may be used as given by

$$\frac{dz}{dt} = Az + \sum_j u_j B^j z + Cu. \tag{5.164}$$

Such a model, as given by Equations (5.162)–(5.164), is a forward model and the parameters must be chosen so that the predictions of the model match the data as closely as possible. This may be done by converting the forward model to an observation model by adding an error term. The set of parameters to be estimated, $\vec{\theta}$, may be partitioned into $\vec{\theta}^h$, the hemodynamic response parameters, and $\vec{\theta}^c$, the connectivity parameters. Prior probabilities for $\vec{\theta}^c$ may be imposed by solution stability requirements and prior probabilities for the parameters $\vec{\theta}^h$ may be imposed by previous measurements of a given individual using simple experimental protocols aimed at measuring the hemodynamic parameters. Putting the prior probabilities together with the observation model yields a posterior probability Bayesian model that may be estimated using an EM algorithm.

Given the lack of detailed knowledge of the neural networks underlying many cognitive processes, especially higher processes involving language where there are no animal models, several DCMs may be applied to a given analysis problem. These models may be compared to each other using Bayes factors to determine whether the evidence favors one model or the other [358]. Bayes factors are based on the model evidence which is defined as

$$p(y|m) = \int p(y|\theta, m) \, p(\theta|m) \, d\theta, \tag{5.165}$$

where m denotes the model, y denotes the data and θ denotes the model parameters. The Bayes factor for comparing model i to model j is given by

$$B_{ij} = \frac{p(y|m = i)}{p(y|m = j)}. \tag{5.166}$$

The (logarithm of the) model evidence may be approximated by the Bayesian information criterion (BIC) or by the Akaike information criterion (AIC) given by

$$\text{BIC} = \text{Accuracy}(m) - \frac{p}{2} \ln(N_s), \tag{5.167}$$

$$\text{AIC} = \text{Accuracy}(m) - p, \tag{5.168}$$

where Accuracy(m) is a function of error covariances, p is the number of parameters in the model and N_s is the number of scans in the fMRI data set. The BIC is biased toward simple models and the AIC is biased toward complex models so Penny *et al.* [358] recommend computing Bayes factors using both the BIC and the AIC and to make a decision about whether to use one model over another only if the two Bayes factor estimates agree.

6

Conclusion

There is no one best way to reduce fMRI data to brain activity maps, either activation maps or connectivity maps. The fMRI investigator must select a method of choice on the basis of a number of factors including software availability, speed of computation, ease of use and taste. The factor of taste can only have legitimacy if the investigator understands how each of the various approaches work and what the trade-offs are of using one method over another. The goal of this review was to give the investigator that overview of how the various available methods work. Once a method is selected, a deeper understanding of it may be obtained from the original literature. At that point investigators may be confident that, in focusing on a small number of methods for use in their laboratory or clinic, they have not overlooked a method that may have significant impact on the interpretations of their computed brain activity maps.

All of the methods reviewed here require that the MRI data be transferred "off-line" from the MRI computer to a computer dedicated to brain map computation. The only method widely available "on-line" for the computation of activation maps is a simple independent samples t-test that compares average activation in an "on" state to the average activation in an "off" state. Currently, setting up an fMRI capacity in a clinical situation is not a turn-key proposition, although a few turn-key off-line systems are now being offered commercially. The use of even these turn-key off-line systems requires a team of people who between them understand the MRI physics, the cognitive neuropsychology, the neuroanatomy and the brain map computation methods to correctly interpret the computed brain activity maps. This complexity in the interpretation of brain activity maps has led many clinical fMRI ventures to become closely associated with a research center.

For clinical use, as a rule, activation maps only are used. Connectivity mapping is not routinely applied in the clinic. Much more needs to be understood about the physiology of the networked neural-hemodynamic system that is the brain before clinical level confidence may be placed in a dynamic causal model, for example. However, from the fundamental knowledge point of view, effective connectivity modeling promises to provide unique windows into the understanding of how the brain and, ultimately we hope, how the mind works.

References

[1] Adler R. *The Geometry of Random Fields*. New York, Wiley. 1981.

[2] Aguirre G. K., Zarahn E., D'Esposito M. Empirical analyses of BOLD fMRI statistics. II. Spatially smoothed data collected under null-hypothesis and experimental conditions. *Neuroimage*. 1997; **5**:199–212.

[3] Aguirre G. K., Zarahn E., D'Esposito M. The inferential impact of global covariates in functional neuroimaging analysis. *Neuroimage*. 1998; **8**:302–6.

[4] Aguirre G. K., Zarahn E., D'Esposito M. The variability of human, BOLD hemodynamic responses. *Neuroimage*. 1998; **8**:360–9.

[5] Aguirre G. K., Zarahn E., D'Esposito M. A critique of the use of the Kolmogorov-Smirnov (KS) statistic for the analysis of BOLD fMRI data. *Magn Reson Med*. 1998; **39**:500–5.

[6] Ahn C. B., Kim J. H., Cho Z. H. High-speed spiral-scan echo planar NMR imaging – I. *IEEE Trans Med Imag*. 1986; **5**:2–7.

[7] Alexander M. E. Fast hierarchical noniterative registration algorithm. *Int J Imaging Sys Technol*. 1999; **10**:242–57.

[8] Alexander M. E., Baumgartner R., Summers A. R., Windischberger C., Klarhoefer M., Moser E., Somorjai R. L. A wavelet-based method for improving signal-to-noise ratio and contrast in MR images. *Magn Reson Imaging*. 2000; **18**:169–80.

[9] Alexander M. E., Baumgartner R., Windischberger C., Moser E., Somorjai R. L. Wavelet domain de-noising of time-courses in MR image sequences. *Magn Reson Imaging*. 2000; **18**:1129–34.

[10] Almeida R., Ledberg A. Exact multivariate tests for brain imaging data. *Hum Brain Mapp*. 2002; **16**:24–35.

[11] Andersen A. H., Gash D. M., Avison M. J. Principal component analysis of the dynamic response measured by fMRI: a generalized linear systems framework. *Magn Reson Imaging*. 1999; **17**:795–815.

[12] Andersen A. H., Rayens W. S. Structure-seeking multilinear methods for the analysis of fMRI data. *Neuroimage*. 2004; **22**:728–39.

[13] Andersson J. L., Hutton C., Ashburner J., Turner R., Friston K. Modeling geometric deformations in EPI time series. *Neuroimage*. 2001; **13**:903–19.

[14] Andrade A., Kherif F., Mangin J. F., Worsley K. J., Paradis A. L., Simon O., Dehaene S., Le Bihan D., Poline J. B. Detection of fMRI activation using cortical surface mapping. *Hum Brain Mapp*. 2001; **12**:79–93.

[15] Ardekani B. A., Kanno I. Statistical methods for detecting activated regions in functional MRI of the brain. *Magn Reson Imaging*. 1998; **16**:1217–25.

[16] Ardekani B. A., Kershaw J., Kashikura K., Kanno I. Activation detection in functional MRI using subspace modeling and maximum likelihood estimation. *IEEE Trans Med Imag*. 1999; **18**:101–14.

[17] Arfanakis K., Cordes D., Haughton V. M., Moritz C. H., Quigley M. A., Meyerand M. E. Combining independent component analysis and correlation analysis

to probe interregional connectivity in fMRI task activation datasets. *Magn Reson Imaging*. 2000; **18**:921–30.

[18] Ashburner J., Friston K. Multimodal image coregistration and partitioning – a unified framework. *Neuroimage*. 1997; **6**:209–17.

[19] Ashburner J., Andersson J. L., Friston K. J. High-dimensional image registration using symmetric priors. *Neuroimage*. 1999; **9**:619–28.

[20] Ashburner J., Friston K. J. Nonlinear spatial normalization using basis functions. *Hum Brain Mapp*. 1999; **7**:254–66.

[21] Ashburner J., Andersson J. L., Friston K. J. Image registration using a symmetric prior–in three dimensions. *Hum Brain Mapp*. 2000; **9**:212–25.

[22] Aubert A., Costalat R. A model of the coupling between brain electrical activity, metabolism, and hemodynamics: application to the interpretation of functional neuroimaging. *Neuroimage*. 2002; **17**:1162–81.

[23] d'Avossa G., Shulman G. L., Corbetta M. Identification of cerebral networks by classification of the shape of BOLD responses. *J Neurophysiol*. 2003; **90**:360–71. Epub 2003 Mar 26.

[24] Backfrieder W., Baumgartner R., Samal M., Moser E., Bergmann H. Quantification of intensity variations in functional MR images using rotated principal components. *Phys Med Biol*. 1996; **41**:1425–38.

[25] Bagarinao E., Matsuo K., Nakai T., Sato S. Estimation of general linear model coefficients for real-time application. *Neuroimage*. 2003; **19**:422–9.

[26] Bandettini P. A., Jesmanowicz A., Wong E. C., Hyde J. S. Processing strategies for time-course data sets in functional MRI of the human brain. *Magn Reson Med*. 1993; **30**:161–73.

[27] Bandettini P. A., Cox R. W. Event-related fMRI contrast when using constant interstimulus interval: theory and experiment. *Magn Reson Med*. 2000; **43**:540–8.

[28] Barth M., Diemling M., Moser E. Modulation of signal changes in gradient-recalled echo functional MRI with increasing echo time correlate with model calculations. *Magn Reson Imaging*. 1997; **15**:745–52.

[29] Barth M., Windischberger C., Klarhofer M., Moser E. Characterization of BOLD activation in multi-echo fMRI data using fuzzy cluster analysis and a comparison with quantitative modeling. *NMR Biomed*. 2001; **14**:484–9.

[30] Baumgartner R., Scarth G., Teichtmeister C., Somorjai R., Moser E. Fuzzy clustering of gradient-echo functional MRI in the human visual cortex. Part I: reproducibility. *J Magn Reson Imaging*. 1997; **7**:1094–101.

[31] Baumgartner R., Windischberger C., Moser E. Quantification in functional magnetic resonance imaging: fuzzy clustering vs. correlation analysis. *Magn Reson Imaging*. 1998; **16**:115–25.

[32] Baumgartner R., Somorjai R., Summers R., Richter W. Assessment of cluster homogeneity in fMRI data using Kendall's coefficient of concordance. *Magn Reson Imaging*. 1999; **17**:1525–32.

[33] Baumgartner R., Ryner L., Richter W., Summers R., Jarmasz M., Somorjai R. Comparison of two exploratory data analysis methods for fMRI: fuzzy clustering vs. principal component analysis. *Magn Reson Imaging*. 2000; **18**:89–94.

[34] Baumgartner R., Somorjai R., Summers R., Richter W., Ryner L. Correlator beware: correlation has limited selectivity for fMRI data analysis. *Neuroimage*. 2000; **12**:240–3.

[35] Baumgartner R., Somorjai R., Summers R., Richter W., Ryner L., Jarmasz M. Resampling as a cluster validation technique in fMRI. *J Magn Reson Imaging*. 2000; **11**:228–31.

[36] Baumgartner R., Somorjai R. Graphical display of fMRI data: visualizing multidimensional space. *Magn Reson Imaging*. 2001; **19**:283–6.

[37] Baune A., Sommer F. T., Erb M., Wildgruber D., Kardatzki B., Palm G., Grodd W. Dynamical cluster analysis of cortical fMRI activation. *Neuroimage*. 1999; **9**:477–89.

[38] Beckmann C. F., Jenkinson M., Smith S. M. General multilevel linear modeling for group analysis in FMRI. *Neuroimage*. 2003; **20**:1052–63.

[39] Bellgowan P. S. F., Zaad Z. S., Bandettini P. A. Understanding neural system dynamics through task modulation and measurement of functional MRI amplitude, latency, and width. *Proc Natl Acad Sci USA*. 2003; **100**:1415–9.

[40] Benardete E. A., Victor J. D. An extension of the m-sequence technique for the analysis of multi-input nonlinear systems. In: *Advanced Methods of Physiological System Modeling*, Vol. III. Marmarelis V. Z., Ed, New York, Plenum. 1994; pp. 87–110.

[41] Bernardo J. M., Smith A. F. M. *Bayesian Theory*. New York, Wiley. 2000.

[42] Birn R. M., Bandettini P. A., Cox R. W., Jesmanowicz A., Shaker R. Magnetic field changes in the human brain due to swallowing or speaking. *Magn Reson Med*. 1998; **40**:55–60.

[43] Birn R. M., Bandettini P. A., Cox R. W., Shaker R. Event-related fMRI of tasks involving brief motion. *Hum Brain Mapp*. 1999; **7**:106–14.

[44] Birn R. M., Cox R. W., Bandettini P. A. Detection versus estimation in event-related fMRI: choosing the optimal stimulus timing. *Neuroimage*. 2002; **15**:252–64.

[45] Biswal B., Yetkin F. Z., Haughton V. M., Hyde J. S. Functional connectivity in the motor cortex of resting human brain using echo-planar MRI. *Magn Reson Med*. 1995; **34**:537–41.

[46] Biswal B., DeYoe A. E., Hyde J. S. Reduction of physiological fluctuations in fMRI using digital filters. *Magn Reson Med*. 1996; **35**:107–13.

[47] Biswal B. B., Hyde J. S. Contour-based registration technique to differentiate between task-activated and head motion-induced signal variations in fMRI. *Magn Reson Med*. 1997; **38**:470–6.

[48] Biswal B. B., Van Kylen J., Hyde J. S. Simultaneous assessment of flow and BOLD signals in resting-state functional connectivity maps. *NMR Biomed*. 1997; **10**:165–70.

[49] Biswal B. B., Ulmer J. L. Blind source separation of multiple signal sources of fMRI data sets using independent component analysis. *J Comput Assist Tomogr*. 1999; **23**:265–71.

[50] Bodurka J., Jesmanowicz A., Hyde J. S., Xu H., Estkowski L., Li S. J. Current-induced magnetic resonance phase imaging. *J Magn Reson*. 1999; **137**:265–71.

[51] Bodurka J., Bandettini P. A. Toward direct mapping of neuronal activity: MRI detection of ultraweak, transient magnetic field changes. *Magn Reson Med*. 2002; **47**:1052–1058.

[52] Box G. E. P., Jenkins G. M., Reinsel G. C. *Time Series Analysis: Forecasting and Control*. Englewood Cliffs, Prentice Hall. 1994.

[53] Boxerman J. L., Bandettini A., Kwong K. K., Baker J. R., Davis T. L., Rosen B. R., Weisskoff R. M. The intravascular contribution to fMRI signal change: Monte Carlo modeling and diffusion-weighted studies in vivo. *Magn Reson Med*. 1995; **32**: 749–63.

[54] Boynton G. M., Engel S. A., Glover G. H., Heeger D. J. Linear systems analysis of functional magnetic resonance imaging in human V1. *J Neurosci*. 1996; **16**: 4207–21.

[55] Brammer M. J. Multidimensional wavelet analysis of functional magnetic resonance images. *Hum Brain Mapp*. 1998; **6**:378–82.

[56] Breakspear M., Brammer M. J., Bullmore E. T., Das P., Williams L. M. Spatiotemporal wavelet resampling for functional neuroimaging data. *Hum Brain Mapp.* 2004; **23**:1–25.

[57] Brigham E. O. *The Fast Fourier Transform.* Englewood Cliffs, Prentice-Hall. 1974.

[58] Büchel C., Friston K. J. Modulation of connectivity in visual pathways by attention: cortical interactions evaluated with structural equation modeling and fMRI. *Cereb Cortex.* 1997; **7**:768–78.

[59] Büchel C., Friston K. J. Dynamic changes in effective connectivity characterized by variable parameter regression and Kalman filtering. *Hum Brain Mapp.* 1998; **6**:403–8.

[60] Büchel C., Holmes A. P., Rees G., Friston K. J. Characterizing stimulus-response functions using nonlinear regressors in parametric fMRI experiments. *Neuroimage.* 1998; **8**:140–8.

[61] Büchel C., Coull J. T., Friston K. J. The predictive value of changes in effective connectivity for human learning. *Science.* 1999; **283**:1538–41.

[62] Büchel C., Friston K. Assessing interactions among neuronal systems using functional neuroimaging. *Neural Netw.* 2000; **13**:871–82.

[63] Büchel C., Friston K. Interactions among neuronal systems assessed with functional neuroimaging. *Rev Neurol (Paris).* 2001; **157**:807–15.

[64] Buckner R. L. Event-related fMRI and the hemodynamic response. *Hum Brain Mapp.* 1998; **6**:373–7.

[65] Buckner R. L., Bandettini P. A., O'Craven K. M., Savoy R. L., Petersen S. E., Raichle M. E., Rosen B. R. Detection of cortical activation during averaged single trials of a cognitive task using functional magnetic resonance. *Proc Natl Acad Sci USA.* 1996; **93**:14878–83.

[66] Buckner R. L. The hemodynamic inverse problem: making inferences about neural activity from measured MRI signals. *Proc Natl Acad Sci USA.* 2003; **100**:2177–9.

[67] Bullmore E. T., Brammer M. J., Williams S. C. R., Rabe-Hesketh S., Janot N., David A., Mellers J., Howard R., Sham P. Statistical methods of estimation and inference for functional MR images. *Magn Reson Med.* 1996; **35**:261–77.

[68] Bullmore E. T., Rabe-Hesketh S., Morris R. G., Williams S. C., Gregory L., Gray J. A., Brammer M. J. Functional magnetic resonance image analysis of a large-scale neurocognitive network. *Neuroimage.* 1996; **4**:16–33.

[69] Bullmore E., Long C., Suckling J., Fadili J., Calvert G., Zelaya F., Carpenter T. A., Brammer M. Colored noise and computational inference in neurophysiological (fMRI) time series analysis: resampling methods in time and wavelet domains. *Hum Brain Mapp.* 2001; **12**:61–78.

[70] Bullmore E., Fadili J., Breakspear M., Salvador R., Suckling J., Brammer M. Wavelets and statistical analysis of functional magnetic resonance images of the human brain. *Stat Methods Med Res.* 2003; **12**:375–99.

[71] Buonocore M. H., Gao L. Ghost artifact reduction for echo planar imaging using phase correction. *Magn Reson Med.* 1997; **38**:89–100.

[72] Buonocore M. H., Zhu D. C. Image-based ghost correction for interleaved EPI. *Magn Reson Med.* 2001; **45**:96–108.

[73] Buracas G. T., Boynton G. M. Efficient design of event-related fMRI experiments using m-sequences. *Neuroimage.* 2002; **16**:801–813.

[74] Burock M. A., Buckner R. L., Woldorff M. G., Rosen B. R., Dale A. M. Randomized event-related experimental designs allow for extremely rapid presentation rates using functional MRI. *Neuroreport.* 1998; **9**:3735–9.

[75] Buxton R. B., Frank L. R. A model for the coupling between cerebral blood flow and oxygen metabolism during neural stimulation. *J Cereb Blood Flow Metab*. 1997; **17**:64–72.

[76] Buxton R. B., Wong E. C., Frank L. R. Dynamics of blood flow and oxygenation changes during brain activation: the balloon model. *Magn Reson Med*. 1998; **39**:855–64.

[77] Buxton R. B. The elusive initial dip. *Neuroimage*. 2001; **13**:953–8.

[78] Calhoun V., Adali T., Kraut M., Pearlson G. A weighted least-squares algorithm for estimation and visualization of relative latencies in event-related functional MRI. *Magn Reson Med*. 2000; **44**:947–54.

[79] Calhoun V. D., Adali T., Pearlson G. D., Pekar J. J. Spatial and temporal independent component analysis of functional MRI data containing a pair of task-related waveforms. *Hum Brain Mapp*. 2001; **13**:43–53.

[80] Calhoun V. D., Adali T., Pearlson G. D., Pekar J. J. A method for making group inferences from functional MRI data using independent component analysis. *Hum Brain Mapp*. 2001; **14**:140–51. Erratum in: *Hum Brain Mapp*. 2002; **16**:131.

[81] Calhoun V. D., Adali T., McGinty V. B., Pekar J. J., Watson T. D., Pearlson G. D. fMRI activation in a visual-perception task: network of areas detected using the general linear model and independent components analysis. *Neuroimage*. 2001; **14**: 1080–8.

[82] Calhoun V. D., Adali T., Pearlson G. D., van Zijl P. C., Pekar J. J. Independent component analysis of fMRI data in the complex domain. *Magn Reson Med*. 2002; **48**:180–92.

[83] Calhoun V. D., Adali T., Pekar J. J., Pearlson G. D. Latency (in)sensitive ICA. Group independent component analysis of fMRI data in the temporal frequency domain. *Neuroimage*. 2003; **20**:1661–9.

[84] Chen N. K., Wyrwicz A. M. Correction for EPI distortions using multi-echo gradient-echo imaging. *Magn Reson Med*. 1999; **41**:1206–13.

[85] Chen N. K., Wyrwicz A. M. Optimized distortion correction technique for echo planar imaging. *Magn Reson Med*. 2001; **45**:525–8.

[86] Chen S., Bouman C. A., Lowe M. J. Clustered components analysis for functional MRI. *IEEE Trans Med Imag*. 2004; **23**:85–98.

[87] Chen H., Yao D. Discussion on the choice of separated components in fMRI data analysis by spatial independent component analysis. *Magn Reson Imaging*. 2004; **22**:827–33.

[88] Chiou J. Y., Ahn C. B., Muftuler L. T., Nalcioglu O. A simple simultaneous geometric and intensity correction method for echo-planar imaging by EPI-based phase modulation. *IEEE Trans Med Imaging*. 2003; **22**:200–5.

[89] Cho Z. H., Park S. H., Kim J. H., Chung S. C., Chung S. T., Chung J. Y., Moon C. W., Yi J. H., Sin C. H., Wong E. K. Analysis of acoustic noise in MRI. *Magn Reson Imaging*. 1997; **15**:815–22.

[90] Cho Z. H., Chung S. T., Chung J. Y., Park S. H., Kim J. S., Moon C. H., Hong I. K. A new silent magnetic resonance imaging using a rotating DC gradient. *Magn Reson Med*. 1998; **39**:317–21.

[91] Cho Z. H., Chung S. C., Lim D. W., Wong E. K. Effects of the acoustic noise of the gradient systems on fMRI: a study on auditory, motor, and visual cortices. *Magn Reson Med*. 1998; **39**:331–6.

[92] Chuang K. H., Chiu M. J., Lin C. C., Chen J. H. Model-free functional MRI analysis using Kohonen clustering neural network and fuzzy C-means. *IEEE Trans Med Imag*. 1999; **18**:1117–28.

[93] Chuang K. H., Chen J. H. IMPACT: image-based physiological artifacts estimation and correction technique for functional MRI. *Magn Reson Med*. 2001; **46**:344–53.

[94] Chui H., Win L., Schultz R., Duncan J. S., Rangarajan A. A unified non-rigid feature registration method for brain mapping. *Med Image Anal*. 2003; **7**:113–30.

[95] Ciulla C., Deek F. P. Development and characterization of an automatic technique for the alignment of fMRI time series. *Brain Topogr*. 2001; **14**:41–56.

[96] Ciulla C., Deek F. P. Performance assessment of an algorithm for the alignment of fMRI time series. *Brain Topogr*. 2002; **14**:313–32.

[97] Clare S., Humberstone M., Hykin J., Blumhardt L. D., Bowtell R., Morris P. Detecting activations in event-related fMRI using analysis of variance. *Magn Reson Med*. 1999; **42**:1117–22.

[98] Clark V. P. Orthogonal polynomial regression for the detection of response variability in event-related fMRI. *Neuroimage*. 2002; **17**:344–63.

[99] Cohen M. S. Parametric analysis of fMRI data using linear systems methods. *Neuroimage*. 1997; **6**:93–103.

[100] Collins D., Peters T., Evans A. An automated 3D nonlinear image deformation procedure for determination of gross morphometric variability in human brain. *Proc Visual Biomed Comput*. 1992; **2359**:541–51.

[101] Comon P. Independent component analysis: A new concept? *Signal Processing*. 1994; **36**:11–20.

[102] Constable R. T., Spencer D. D. Repetition time in echo planar functional MRI. *Magn Reson Med*. 2001; **46**:748–55.

[103] Cordes D., Haughton V. M., Arfanakis K., Wendt G. J., Turski P. A., Moritz C. H., Quigley M. A., Meyerand M. E. Mapping functionally related regions of brain with functional connectivity MR imaging. *Am J Neuroradiol*. 2000; **21**:1636–44.

[104] Cordes D., Haughton V., Carew J. D., Arfanakis K., Maravilla K. Hierarchical clustering to measure connectivity in fMRI resting-state data. *Magn Reson Imaging*. 2002; **20**:305–17.

[105] Corfield D. R., Murphy K., Josephs O., Fink G. R., Frackowiak R. S., Guz A., Adams L., Turner R. Cortical and subcortical control of tongue movement in humans. *J Appl Physiol*. 1999; **86**:1468–77.

[106] Cox D. D., Savoy R. L. Functional magnetic resonance imaging (fMRI) "brain reading": detecting and classifying distributed patterns of fMRI activity in human visual cortex. *Neuroimage*. 2003; **19**:261–70.

[107] Cox R. W. AFNI: software for analysis and visualization of functional magnetic resonance neuroimages. *Comput Biomed Res*. 1996; **29**:162–73.

[108] Cox R. W., Jesmanowicz A., Hyde J. S. Real-time functional magnetic resonance imaging. *Magn Reson Med*. 1995; **33**:230–6.

[109] Cox R. W., Hyde J. S. Software tools for analysis and visualization of fMRI data. *NMR Biomed*. 1997; **10**:171–8.

[110] Cox R. W., Jesmanowicz A. Real-time 3D image registration for functional MRI. *Magn Reson Med*. 1999; **42**:1014–8.

[111] Cusack R., Brett M., Osswald K. An evaluation of the use of magnetic field maps to undistort echo-planar images. *Neuroimage*. 2003; **18**:127–42.

[112] Dagli M. S., Ingeholm J. E., Haxby J. V. Localization of cardiac-induced signal change in fMRI. *Neuroimage*. 1999; **9**:407–15.

[113] Dale A. M., Buckner R. L. Selective averaging of rapidly presented individual trials using fMRI. *Hum. Brain. Mapp*. 1997; **5**:329–40.

[114] Daubechies I. Orthonormal bases of compactly supported wavelets. *Comm Pure Appl Math*. 1988; **41**:909–95.

[115] Daubechies I. *Ten Lectures on Wavelets*. Philadelphia, SIAM. 1992.

[116] Davatzikos C. Mapping image data to stereotaxic spaces: applications to brain mapping. *Hum Brain Mapp*. 1998; **6**:334–8.

[117] Deco G., Rolls E. T., Horwitz B. "What" and "where" in visual working memory: a computational neurodynamical perspective for integrating FMRI and single-neuron data. *J Cogn Neurosci*. 2004; **16**:683–701.

[118] Desco M., Hernandez J. A., Santos A., Brammer M. Multiresolution analysis in fMRI: sensitivity and specificity in the detection of brain activation. *Hum Brain Mapp*. 2001; **14**:16–27.

[119] Descombes X., Kruggel F., von Cramon D. Y. fMRI signal restoration using a spatio-temporal Markov random field preserving transitions. *Neuroimage*. 1998; **8**:340–9.

[120] Descombes X., Kruggel F., von Cramon D. Y. Spatio-temporal fMRI analysis using Markov random fields. *IEEE Trans Med Imag*. 1998; **17**:1028–39.

[121] Desjardins A. E., Kiehl K. A., Liddle P. F. Removal of confounding effects of global signal in functional MRI analyses. *Neuroimage*. 2001; **13**:751–8.

[122] Dinov I. D., Mega M. S., Thompson P. M., Woods R. P., Sumners de W. L., Sowell E. L., Toga A. W. Quantitative comparison and analysis of brain image registration using frequency-adaptive wavelet shrinkage. *IEEE Trans Inf Technol Biomed*. 2002; **6**:73–85.

[123] Donoho D. L., Johnstone I. M. Wavelet Shrinkage: Asymptopia? *J Roy Stat Soc*. 1995; **57**:301–69.

[124] Du Y. P., Joe Zhou X., Bernstein M. A. Correction of concomitant magnetic field-induced image artifacts in nonaxial echo-planar imaging. *Magn Reson Med*. 2002; **48**:509–15.

[125] Duann J. R., Jung T. P., Kuo W. J., Yeh T. C., Makeig S., Hsieh J. C., Sejnowski T. J. Single-trial variability in event-related BOLD signals. *Neuroimage*. 2002; **15**:823–35.

[126] Duong T. Q., Kim D. S., Ugurbil K., Kim S. G. Spatiotemporal dynamics of the BOLD fMRI signals: toward mapping submillimeter cortical columns using the early negative response. *Magn Reson Med*. 2000; **44**:231–42.

[127] Duong T. Q., Yacoub E., Adriany G., Hu X., Ugurbil K., Kim S. G. Microvascular BOLD contribution at 4 and 7 T in the human brain: gradient-echo and spin-echo fMRI with suppression of blood effects. *Magn Reson Med*. 2003; **49**:1019–27.

[128] Eddy W. F., Fitzgerald M., Noll D. C. Improved image registration by using Fourier interpolation. *Magn Reson Med*. 1996; **36**:923–31.

[129] Edelstein W. A., Kidane T. K., Taracila V., Baig T. N., Eagan T. P., Cheng Y.-C. N., Brown R. W., Mallick J. A. Active-passive gradient shielding for MRI acoustic noise reduction. *Magn Reson Med*. 2005; **53**:1013–17.

[130] Esposito F., Formisano E., Seifritz E., Goebel R., Morrone R., Tedeschi G., Di Salle F. Spatial independent component analysis of functional MRI time-series: to what extent do results depend on the algorithm used? *Hum Brain Mapp*. 2002; **16**:146–57.

[131] Evans A. C., Collins D. L., Milner B. An MRI-based stereotactic atlas from 250 young normal subjects. *Soc Neurosci Abstr*. 1992; **18**:408.

[132] Evans A. C., Collins D. L., Mills S. R., Brown E. D., Kelly R. L., Peters T. M. 3D statistical neuroanatomical models from 305 MRI volumes. In: *Proceedings of the IEEE Nuclear Science Symposium and Medical Imaging Conference*. 1993; pp. 1813–17.

[133] Evans A. C., Kamber M., Collins D. L., Macdonald D. An MRI-based probabilistic atlas of neuroanatomy. In *Magnetic Resonance Scanning and Epilepsy*. Shorvon S., Fish D., Andermann F., Bydder G. M., Stefan H., Eds., NATO ASI Series A, Life Sciences, Vol. 264. New York, Plenum. 1994.

[134] Fadili M. J., Ruan S., Bloyet D., Mazoyer B. A multistep unsupervised fuzzy clustering analysis of fMRI time series. *Hum Brain Mapp.* 2000; **10**:160–78.

[135] Fadili M. J., Ruan S., Bloyet D., Mazoyer B. On the number of clusters and the fuzziness index for unsupervised FCA application to BOLD fMRI time series. *Med Image Anal.* 2001; **5**:55–67.

[136] Fadili M. J., Bullmore E. T. Wavelet-generalized least squares: a new BLU estimator of linear regression models with 1/f errors. *Neuroimage.* 2002; **15**:217–32.

[137] Feng C. M., Liu H. L., Fox P. T., Gao J. H. Comparison of the experimental BOLD signal change in event-related fMRI with the balloon model. *NMR Biomed.* 2001; **14**:397–401.

[138] Filzmoser P., Baumgartner R., Moser E. A hierarchical clustering method for analyzing functional MR images. *Magn Reson Imaging.* 1999; **17**:817–26.

[139] Fischer H., Hennig J. Neural network-based analysis of MR time series. *Magn Reson Med.* 1999; **41**:124–31.

[140] Fisher R. A. *Statistical Methods for Research Workers.* London, Oliver and Boyd. 1950.

[141] Forman S. D., Cohen J. D., Fitzgerald M., Eddy W. F., Mintun M. A., Noll D. C. Improved assessment of significant activation in functional magnetic resonance imaging (fMRI): use of a cluster-size threshold. *Magn Reson Med.* 1995; **33**:636–47.

[142] Frackowiak R. S. J., Friston K. J., Frith C. D., Dolan R. J., Mazziotta J. *Human Brain Function.* San Diego, Academic Press 1997.

[143] Frank L. R., Buxton R. B., Wong E. C. Probabilistic analysis of functional magnetic resonance imaging data. *Magn Reson Med.* 1998; **39**:132–48.

[144] Frank L. R., Buxton R. B., Wong E. C. Estimation of respiration-induced noise fluctuations from undersampled multislice fMRI data. *Magn Reson Med.* 2001; **45**:635–44.

[145] Friman O., Borga M., Lundberg P., Knutsson H. Exploratory fMRI analysis by autocorrelation maximization. *Neuroimage.* 2002; **16**:454–64.

[146] Friman O., Borga M., Lundberg P., Knutsson H. Adaptive analysis of fMRI data. *Neuroimage.* 2003; **19**:837–45.

[147] Friston K. J. Functional and effective connectivity in neuroimaging: a synthesis. *Hum Brain Mapp.* 1994; **2**:56–78.

[148] Friston K. J. Bayesian estimation of dynamical systems: an application to fMRI. *Neuroimage.* 2002; **16**:513–30.

[149] Friston K. J. Beyond phrenology: What can neuroimaging tell us about distributed circuitry? *Ann Rev Neurosci.* 2002; **25**:221–50.

[150] Friston K. J., Frith C. D., Liddle P., Frackowiak R. S. Comparing functional (PET) images: the assessment of significant change. *J Cereb Blood Flow Metab.* 1991; **11**:690–9.

[151] Friston K. J., Jezzard P., Turner R. Analysis of functional MRI time-series. *Hum Brain Mapp.* 1994; **1**:153–71.

[152] Friston K. J., Worsley K. J., Frackowiak R. S., Mazziotta J., Evans A. C. Assessing the significance of focal activations using their spatial extent. *Hum Brain Mapp.* 1994; **1**:214–20.

[153] Friston K. J., Ashburner J., Frith C. D., Poline J.-B., Heather J. D., Frackowiak R. S. J. Spatial registration and normalization of images. *Hum Brain Mapp.* 1995; **2**:165–89.

[154] Friston K. J., Holmes A. P., Poline J. B., Grasby P. J., Williams S. C., Frackowiak R. S., Turner R. Analysis of fMRI time-series revisited. *Neuroimage.* 1995; **2**:45–53.

[155] Friston K. J., Frith C. D., Turner R., Frackowiak R. S. Characterizing evoked hemodynamics with fMRI. *Neuroimage.* 1995; **2**:157–65.

[156] Friston K. J., Frith C. D., Frackowiak R. S., Turner R. Characterizing dynamic brain responses with fMRI: a multivariate approach. *Neuroimage*. 1995; **2**: 166–72.

[157] Friston K. J., Holmes A., Worsley K. J., Poline J.-P., Frith C. D., Frackowiak RSJ. Statistical parametric mapping: a general linear approach. *Hum Brain Mapp*. 1995; **2**:189–210.

[158] Friston K. J., Holmes A., Poline J. B., Price C. J., Frith C. D. Detecting activations in PET and fMRI: levels of inference and power. *Neuroimage*. 1996; **4**:223–35.

[159] Friston K. J., Frith C. D., Fletcher P., Liddle P. F., Frackowiak R. S. J. Functional topography: Multidimensional scaling and functional connectivity in the brain. *Cereb Cortex*. 1996; **6**:156–64.

[160] Friston K. J., Williams S., Howard R., Frackowiak R. S., Turner R. Movement-related effects in fMRI time-series. *Magn Reson Med*. 1996; **35**:346–55.

[161] Friston K. J., Fletcher P., Josephs O., Holmes A., Rugg M. D., Turner R. Event-related fMRI: characterizing differential responses. *Neuroimage*. 1998; **7**:30–40.

[162] Friston K. J., Josephs O., Rees G., Turner R. Nonlinear event-related responses in fMRI. *Magn Reson Med*. 1998; **39**:41–52.

[163] Friston K. J., Holmes A. P., Price C. J., Buchel C., Worsley K. J. Multisubject fMRI studies and conjunction analyses. *Neuroimage*. 1999; **10**:385–96.

[164] Friston K. J., Zarahn E., Josephs O., Henson R. N., Dale A. M. Stochastic designs in event-related fMRI. *Neuroimage*. 1999; **10**:607–19.

[165] Friston K., Phillips J., Chawla D., Buchel C. Revealing interactions among brain systems with nonlinear PCA. *Hum Brain Mapp*. 1999; **8**:92–7.

[166] Friston K. J., Josephs O., Zarahn E., Holmes A. P., Rouquette S., Poline J.-B. To smooth or not to smooth? Bias and efficiency in fMRI time-series analysis. *Neuroimage*. 2000; **12**:196–208.

[167] Friston K. J., Buchel C. Attentional modulation of effective connectivity from V2 to V5/MT in humans. *Proc Natl Acad Sci USA*. 2000; **97**:7591–6.

[168] Friston K. J., Mechelli A., Turner R., Price C. J. Nonlinear responses in fMRI: the Balloon model, Volterra kernels, and other hemodynamics. *Neuroimage*. 2000; **12**:466–77.

[169] Friston K., Phillips J., Chawla D., Buchel C. Nonlinear PCA: characterizing interactions between modes of brain activity. *Phil Trans R Soc Lond B Biol Sci*. 2000; **355**:135–46.

[170] Friston K. J., Penny W., Phillips C., Kiebel S., Hinton G., Ashburner J. Classical and Bayesian inference in neuroimaging: theory. *Neuroimage*. 2002; **16**:465–83.

[171] Friston K. J., Glaser D. E., Henson R. N., Kiebel S., Phillips C., Ashburner J. Classical and Bayesian inference in neuroimaging: applications. *Neuroimage*. 2002; **16**:484–512.

[172] Friston K. J., Penny W. Posterior probability maps and SPMs. *Neuroimage*. 2003; **19**:1240–9.

[173] Friston K. J., Harrison L., Penny W. Dynamic causal modeling. *Neuroimage*. 2003; **19**:1273–1302.

[174] Gammerman D. *Markov Chain Monte Carlo*. London, Chapman & Hall. 1997.

[175] Gao J. H., Miller I., Lai S., Xiong J., Fox P. T. Quantitative assessment of blood inflow effects in functional MRI signals. *Magn Reson Med*. 1996; **36**:314–319.

[176] Gao J. H., Yee S. H. Iterative temporal clustering analysis for the detection of multiple response peaks in fMRI. *Magn Reson Imaging*. 2003; **21**:51–3.

[177] Gautama T., Mandic D. P., Van Hulle M. M. Signal nonlinearity in fMRI: a comparison between BOLD and MION. *IEEE Trans Med Imag*. 2003; **22**:636–44.

[178] Gavrilescu M., Shaw M. E., Stuart G. W., Eckersley P., Svalbe I. D., Egan G. F. Simulation of the effects of global normalization procedures in functional MRI. *Neuroimage*. 2002; **17**:532–42.

[179] Genovese C. R., Noll D. C., Eddy W. F. Estimating test–retest reliability in functional MR imaging. I: Statistical methodology. *Magn Reson Med*. 1997; **38**:497–507.

[180] Genovese C. R., Lazar N. A., Nichols T. E. Thresholding of statistical maps in functional neuroimaging using the false discovery rate. *Neuroimage*. 2002; **15**:772–86.

[181] Gibbons R. D., Lazar N. A., Bhaumik D. K., Sclove S. L., Chen H. Y., Thulborn K. R., Sweeney J. A., Hur K., Patterson D. Estimation and classification of fMRI hemodynamic response patterns. *Neuroimage*. 2004; **22**:804–14.

[182] Gilks W., Richardson S., Spiegalhalter D. *Markov Chain Monte Carlo in Practice*. London, Chapman & Hall. 1996.

[183] Gitelman D. R., Penny W. D., Ashburner J., Friston K. J. Modeling regional and psychophysiologic interactions in fMRI: the importance of hemodynamic deconvolution. *Neuroimage*. 2003; **19**:200–7.

[184] Glover G. H. Deconvolution of impulse response in event-related BOLD fMRI. *Neuroimage*. 1999; **9**:416–29.

[185] Glover G. H., Li T. Q., Ress D. Image-based method for retrospective correction of physiological motion effects in fMRI: RETROICOR. *Magn Reson Med*. 2000; **44**:162–7.

[186] Godtliebsen F., Chu C. K., Sorbye S. H., Torheim G. An estimator for functional data with application to MRI. *IEEE Trans Med Imag*. 2001; **20**:36–44.

[187] Goebel R., Roebroeck A., Kim D. S., Formisano E. Investigating directed cortical interactions in time-resolved fMRI data using vector autoregressive modeling and Granger causality mapping. *Magn Reson Imaging*. 2003; **21**:1251–61.

[188] Golay X., Kollias S., Stoll G., Meier D., Valavanis A., Boesiger P. A new correlation-based fuzzy logic clustering algorithm for fMRI. *Magn Reson Med*. 1998; **40**:249–60.

[189] Gold S., Christian B., Arndt S., Cizadlo T., Johnson D. L., Flaum M, Andreasen NC. Functional MRI statistical software packages: a comparative analysis. *Hum Brain Mapp*. 1998; **6**:73–84.

[190] Goncalves M. S., Hall D. A. Connectivity analysis with structural equation modeling: an example of the effects of voxel selection. *Neuroimage*. 2003; **20**:1455–67.

[191] Gössl C., Auer D. P., Fahrmeir L. Dynamic models in fMRI. *Magn Reson Med*. 2000; **43**:72–81.

[192] Gossl C., Fahrmeir L., Auer D. P. Bayesian modeling of the hemodynamic response function in BOLD fMRI. *Neuroimage*. 2001; **14**:140–8.

[193] Gossl C., Auer D. P., Fahrmeir L. Bayesian spatiotemporal inference in functional magnetic resonance imaging. *Biometrics*. 2001; **57**:554–62.

[194] Goutte C., Toft P., Rostrup E., Nielsen F., Hansen L. K. On clustering fMRI time series. *Neuroimage*. 1999; **9**:298–310.

[195] Goutte C., Nielsen F. A., Hansen L. K. Modeling the haemodynamic response in fMRI using smooth FIR filters. *IEEE Trans Med Imag*. 2000; **19**:1188–201.

[196] Goutte C., Hansen L. K., Liptrot M. G., Rostrup E. Feature-space clustering for fMRI meta-analysis. *Hum Brain Mapp*. 2001; **13**:165–83.

[197] Greicius M. D., Krasnow B., Reiss A. L., Menon V. Functional connectivity in the resting brain: a network analysis of the default mode hypothesis. *Proc Natl Acad Sci USA*. 2003; **100**:253–8.

[198] Grootoonk S., Hutton C., Ashburner J., Howseman A. M., Josephs O., Rees G., Friston K. J., Turner R. Characterization and correction of interpolation effects in the realignment of fMRI time series. *Neuroimage*. 2000; **11**:49–57.

[199] Guimaraes A. R., Melcher J. R., Talavage T. M., Baker J. R., Rosen B. R., Weisskoff R. M. Detection of inferior colliculus activity during auditory stimulation using cardiac gated functional MRI with T_1 correction. *Neuroimage*. 1996; **3**:S9.

[200] Haacke E. M., Brown R. W., Thompson M. R., Venkatesan R. *Magnetic Resonance Imaging: Physical Principles and Sequence Design*. New York, Wiley-Liss. 1999.

[201] Hagmann P., Thiran J. P., Jonasson L., Vandergheynst P., Clarke S., Maeder P., Meuli R. DTI mapping of human brain connectivity: statistical fibre tracking and virtual dissection. *Neuroimage*. 2003; **19**:545–54.

[202] Hajnal J. V., Myers R., Oatridge A., Schwieso J. E., Young I. R., Bydder G. M. Artifacts due to stimulus correlated motion in functional imaging of the brain. *Magn Reson Med*. 1994; **31**:283–91.

[203] Hajnal J. V., Saeed N., Soar E. J., Oatridge A., Young I. R., Bydder G. M. A registration and interpolation procedure for subvoxel matching of serially acquired MR images. *J Comput Assist Tomogr*. 1995; **19**:289–96.

[204] Hall D. A., Goncalves M. S., Smith S., Jezzard P., Haggard M. P., Kornak J. A method for determining venous contribution to BOLD contrast sensory activation. *Magn Reson Image*. 2002; **20**:695–705.

[205] Hampson M., Peterson B. S., Skudlarski P., Gatenby J. C., Gore J. C. Detection of functional connectivity using temporal correlations in MR images. *Hum Brain Mapp*. 2002; **15**:247–62.

[206] Handwerker D. A., Ollinger J. M., D'Esposito M. Variation of BOLD hemodynamic responses across subjects and brain regions and their effects on statistical analyses. *Neuroimage*. 2004; **21**:1639–51.

[207] Hansen L. K., Nielsen F. A., Toft P., Liptrot M. G., Goutte C., Strother S. C., Lange N., Gade A., Rottenberg D. A., Paulson O. B. "lyngby" – A modeler's matlab toolbox for spatio-temporal analysis of functional neuroimages. *Neuroimage*. 1999; **9**(6):5241.

[208] Hansen L. K., Nielsen F. A., Strother S. C., Lange N. Consensus inference in neuroimaging. *Neuroimage*. 2001; **13**:1212–18.

[209] Hansen L. K., Nielsen F. A., Larsen J. Exploring fMRI data for periodic signal components. *Artif Intell Med*. 2002; **25**:35–44.

[210] Hanson S. J., Bly B. M. The distribution of BOLD susceptibility effects in the brain is non Gaussian. *Neuroreport*. 2001; **12**:1971–7.

[211] Harms M. P., Melcher J. R. Detection and quantification of a wide range of fMRI temporal responses using a physiologically-motivated basis set. *Hum Brain Mapp*. 2003; **20**:168–83.

[212] Harrison L., Penny W. D., Friston K. Multivariate autoregressive modeling of fMRI time series. *Neuroimage*. 2003; **19**:1477–91.

[213] Hathout G. M., Gambhir S. S., Gopi R. K., Kirlew K. A., Choi Y., So G., Gozal D., Harper R., Lufkin R. B., Hawkins R. A quantitative physiologic model of blood oxygenation for functional magnetic resonance imaging. *Invest Radiol*. 1995; **30**:669–82.

[214] Hayasaka S., Nichols T. E. Validating cluster size inference: random field and permutation methods. *Neuroimage*. 2003; **20**:2343–56.

[215] Hayasaka S., Phan K. L., Liberzon I., Worsley K. J., Nichols T. E. Nonstationary cluster-size inference with random field and permutation methods. *Neuroimage*. 2004; **22**:676–87.

[216] Henson R. N., Price C. J., Rugg M. D., Turner R., Friston K. J. Detecting latency differences in event-related BOLD responses: application to words versus nonwords and initial versus repeated face presentations. *Neuroimage*. 2002; **15**:83–97.

[217] Hernandez L., Badre D., Noll D., Jonides J. Temporal sensitivity of event-related fMRI. *Neuroimage*. 2002; **17**:1018–26.

[218] Hilton M., Ogden T., Hattery D., Eden G., Jawerth B. Wavelet denoising of functional MRI data. In: *Wavelets in Medicine and Biology*. Aldroubi A., Unser M., Eds., Boca Raton, CRC Press. 1996; pp. 93–114.

[219] Himberg J., Hyvarinen A., Esposito F. Validating the independent components of neuroimaging time series via clustering and visualization. *Neuroimage*. 2004; **22**:1214–22.

[220] Hinterberger T., Veit R., Strehl U., Trevorrow T., Erb M., Kotchoubey B., Flor H., Birbaumer N. Brain areas activated in fMRI during self-regulation of slow cortical potentials (SCPs). *Exp Brain Res*. 2003; **152**:113–22.

[221] Honey G. D., Fu C. H., Kim J., Brammer M. J., Croudace T. J., Suckling J., Pich E. M., Williams S. C., Bullmore E. T. Effects of verbal working memory load on corticocortical connectivity modeled by path analysis of functional magnetic resonance imaging data. *Neuroimage*. 2002; **17**:573–82.

[222] Hoogenraad F. G., Pouwels P. J., Hofman M. B., Reichenbach J. R., Sprenger M., Haacke E. M. Quantitative differentiation between BOLD models in fMRI. *Magn Reson Med*. 2001; **45**:233–46.

[223] Hopfinger J. B., Buchel C., Holmes A. P., Friston K. J. A study of analysis parameters that influence the sensitivity of event-related fMRI analyses. *Neuroimage*. 2000; **11**:326–33.

[224] Horwitz B. The elusive concept of brain connectivity. *Neuroimage*. 2003; **19**:466–70.

[225] Hossein-Zadeh G. A., Soltanian-Zadeh H., Ardekani B. A. Multiresolution fMRI activation detection using translation invariant wavelet transform and statistical analysis based on resampling. *IEEE Trans Med Imag*. 2003; **22**:302–14.

[226] Hossein-Zadeh G. A., Ardekani B. A., Soltanian-Zadeh H. Activation detection in fMRI using a maximum energy ratio statistic obtained by adaptive spatial filtering. *IEEE Trans Med Imag*. 2003; **22**:795–805.

[227] Howseman A. M., Grootoonk S., Porter D. A., Ramdeen J., Holmes A. P., Turner R. The effect of slice order and thickness on fMRI activation using multislice echo-planar imaging. *Neuroimage* 1999; **9**:363–76.

[228] Hu X., Kim S. G. Reduction of signal fluctuation in functional MRI using navigator echoes. *Magn Reson Med*. 1994; **31**:495–503.

[229] Hu X., Le T. H., Parrish T., Erhard P. Retrospective estimation and correction of physiological fluctuation in functional MRI. *Magn Reson Med*. 1995; **34**:201–12.

[230] Hutton C., Bork A., Josephs O., Deichmann R., Ashburner J., Turner R. Image distortion correction in fMRI: a quantitative evaluation. *Neuroimage*. 2002; **16**:217–40.

[231] Janz C., Heinrich S. P., Kornmayer J., Bach M., Hennig J. Coupling of neural activity and BOLD fMRI response: new insights by combination of fMRI and VEP experiments in transition from single events to continuous stimulation. *Magn Reson Med*. 2001; **46**:482–6.

[232] Jarmasz M., Somorjai R. L. Exploring regions of interest with cluster analysis (EROICA) using a spectral peak statistic for selecting and testing the significance of fMRI activation time-series. *Artif Intell Med*. 2002; **25**:45–67.

[233] Jenkinson M., Smith S. A global optimization method for robust affine registration of brain images. *Med Image Anal*. 2001; **5**:143–56.

[234] Jernigan T. L., Gamst A. C., Fennema-Notestine C., Ostergaard A. L. More "mapping" in brain mapping: statistical comparison of effects. *Hum Brain Mapp*. 2003; **19**:90–95.

[235] Jezzard P., Balaban R. S. Correction for geometric distortion in echo planar images from B_0 field variations. *Magn Reson Med*. 1995; **34**:65–73.

[236] Jezzard P., Clare S. Sources of distortion in functional MRI data. *Hum Brain Mapp.* 1999; **8**:80–5.

[237] Josephs O., Turner R., Friston K. Event-related fMRI. *Hum Brain Mapp.* 1997; **5**:243–8.

[238] Josephs O., Henson R. N. Event-related functional magnetic resonance imaging: modeling, inference and optimization. *Phil Trans R Soc Lond B Biol Sci.* 1999; **354**:1215–28.

[239] Kamba M., Sung Y. W., Ogawa S. A dynamic system model-based technique for functional MRI data analysis. *Neuroimage.* 2004; **22**:179–87.

[240] Katanoda K., Matsuda Y., Sugishita M. A spatio-temporal regression model for the analysis of functional MRI data. *Neuroimage.* 2002; **17**:1415–28.

[241] Kellman P., Gelderen P., de Zwart J. A., Duyn J. H. Method for functional MRI mapping of nonlinear response. *Neuroimage.* 2003; **19**:190–9.

[242] Kennan R. P., Scanley B. E., Innis R. B., Gore J. C. Physiological basis for BOLD MR signal changes due to neuronal stimulation: separation of blood volume and magnetic susceptibility effects. *Magn Reson Med.* 1998; **40**:840–6.

[243] Kershaw J., Ardekani B. A., Kanno I. Application of Bayesian inference to fMRI data analysis. *IEEE Trans Med Imag.* 1999; **18**:1138–53.

[244] Kershaw J., Kashikura K., Zhang X., Abe S., Kanno I. Bayesian technique for investigating linearity in event-related BOLD fMRI. *Magn Reson Med.* 2001; **45**:1081–94.

[245] Kherif F., Poline J. B., Flandin G., Benali H., Simon O., Dehaene S., Worsley K. J. Multivariate model specification for fMRI data. *Neuroimage.* 2002; **16**:1068–83.

[246] Kiebel S. J., Poline J. B., Friston K. J., Holmes A. P., Worsley K. J. Robust smoothness estimation in statistical parametric maps using standardized residuals from the general linear model. *Neuroimage.* 1999; **10**:756–66.

[247] Kiebel S. J., Goebel R., Friston K. J. Anatomically informed basis functions. *Neuroimage.* 2000; **11**:656–67.

[248] Kiebel S., Friston K. J. Anatomically informed basis functions in multisubject studies. *Hum Brain Mapp.* 2002; **16**:36–46.

[249] Kiebel S. J., Glasser D. E., Friston K. J. A heuristic for the degrees of freedom of statistics based on multiple variance parameters. *Neuroimage.* 2003; **20**:591–600.

[250] Kim S. G., Ugurbil K. Comparison of blood oxygenation and cerebral blood flow effects in fMRI: estimation of relative oxygen consumption change. *Magn Reson Med.* 1997; **38**:59–65.

[251] Kim S. G., Ugurbil K. Functional magnetic resonance imaging of the human brain. *J Neurosci Meth.* 1997; **74**:229–43.

[252] Kim D. S., Ronen I., Olman C., Kim S. G., Ugurbil K., Toth L. J. Spatial relationship between neuronal activity and BOLD functional MRI. *Neuroimage.* 2004; **21**:876–85.

[253] Kiviniemi V., Jauhiainen J., Tervonen O., Paakko E., Oikarinen J., Vainionpaa V., Rantala H., Biswal B. Slow vasomotor fluctuation in fMRI of anesthetized child brain. *Magn. Reson. Med.* 2000; **44**:373–378.

[254] Kiviniemi V., Kantola J. H., Jauhiainen J., Hyvarinen A., Tervonen O. Independent component analysis of nondeterministic fMRI signal sources. *Neuroimage.* 2003; **19**:253–60.

[255] Kiviniemi V., Kantola J. H., Jauhiainen J., Tervonen O. Comparison of methods for detecting nondeterministic BOLD fluctuation in fMRI. *Magn Reson Imaging.* 2004; **22**:197–203.

[256] Koch M. A., Norris D. G., Hund-Georgiadis M. An investigation of functional and anatomical connectivity using magnetic resonance imaging. *Neuroimage.* 2002; **16**:241–50.

[257] Kruggel F., von Cramon D. Y., Descombes X. Comparison of filtering methods for fMRI datasets. *Neuroimage.* 1999; **10**:530–43.

[258] Kruggel F., von Cramon D. Y. Modeling the hemodynamic response in single-trial functional MRI experiments. *Magn Reson Med.* 1999; **42**:787–97.

[259] Kruggel F., Pelegrini-Issac M., Benali H. Estimating the effective degrees of freedom in univariate multiple regression analysis. *Med Image Anal.* 2002; **6**:63–75.

[260] LaConte S. M., Ngan S. C., Hu X. Wavelet transform-based Wiener filtering of event-related fMRI data. *Magn Reson Med.* 2000; **44**:746–57.

[261] LaConte S., Anderson J., Muley S., Ashe J., Frutiger S., Rehm K., Hansen L. K., Yacoub E., Hu X., Rottenberg D., Strother S. The evaluation of preprocessing choices in single-subject BOLD fMRI using NPAIRS performance metrics. *Neuroimage.* 2003; **18**:10–27.

[262] Lahaye P. J., Poline J. B., Flandin G., Dodel S., Garnero L. Functional connectivity: studying nonlinear, delayed interactions between BOLD signals. *Neuroimage.* 2003; **20**:962–74.

[263] Lai S. H., Fang M. A novel local PCA-based method for detecting activation signals in fMRI. *Magn Reson Imaging.* 1999; **17**:827–36.

[264] Laird A. R., Rogers B. P., Meyerand M. E. Comparison of Fourier and wavelet resampling methods. *Magn Reson Med.* 2004; **51**:418–22.

[265] Lange N. Statistical thinking in functional and structural magnetic resonance neuroimaging. *Stat Med.* 1999; **18**:2401–7.

[266] Lange N., Zeger S. L. Non-linear Fourier time series analysis for human brain mapping by functional magnetic resonance imaging. *J Roy Stat Soc C App Stat.* 1997; **46**:1–30.

[267] Lange N., Strother S. C., Anderson J. R., Nielsen F. A., Holmes A. P., Kolenda T., Savoy R., Hansen L. K. Plurality and resemblance in fMRI data analysis. *Neuroimage.* 1999; **10**:282–303.

[268] Laurienti P. J., Burdette J. H., Maldjian J. A. Separating neural processes using mixed event-related and epoch-based fMRI paradigms. *J Neurosci Meth.* 2003; **131**:41–50.

[269] Lauterbur P. C. Image formation by induced local interactions: examples employing nuclear magnetic resonance. *Nature.* 1973; **242**:190.

[270] Lazar N. A., Luna B., Sweeney J. A., Eddy W. F. Combining brains: a survey of methods for statistical pooling of information. *Neuroimage.* 2002; **16**:538–50.

[271] Le T. H., Hu X. Retrospective estimation and correction of physiological artifacts in fMRI by direct extraction of physiological activity from MR data. *Magn Reson Med.* 1996; **35**:290–8.

[272] Ledberg A., Fransson P., Larsson J., Petersson K. M. A 4D approach to the analysis of functional brain images: application to FMRI data. *Hum Brain Mapp.* 2001; **13**:185–98.

[273] Lee L., Harrison L. M., Mechelli A. The functional brain connectivity workshop: report and commentary. *Network: Comput Neural Syst.* 2003; **14**:R1-R15.

[274] Lee S. P., Silva A. C., Ugurbil K., Kim S. G. Diffusion-weighted spin-echo fMRI at 9.4 T: microvascular/tissue contribution to BOLD signal changes. *Magn Reson Med.* 1999; **42**:919–28.

[275] Lee S. P., Duong T. Q., Yang G., Iadecola C., Kim S. G. Relative changes of cerebral arterial and venous blood volumes during increased cerebral blood flow: implications for BOLD fMRI. *Magn Reson Med.* 2001; **45**:791–800.

[276] Levin D. N., Uftring S. J. Detecting brain activation in fMRI data without prior knowledge of mental event timing. *Neuroimage* 2000; **13**:153–60.

[277] Liao C. H., Worsley K. J., Poline J. B., Aston J. A., Duncan G. H., Evans A. C. Estimating the delay of the fMRI response. *Neuroimage*. 2002; **16**:593–606.

[278] Lipschutz B., Friston K. J., Ashburner J., Turner R., Price C. J. Assessing study-specific regional variations in fMRI signal. *Neuroimage*. 2001; **13**:392–8.

[279] Liou M., Su H. R., Lee J. D., Cheng P. E., Huang C. C., Tsai C. H. Functional MR images and scientific inference: reproducibility maps. *J Cogn Neurosci*. 2003; **15**:935–45.

[280] Liu T. T. Efficiency, power, and entropy in event-related fMRI with multiple trial types. Part II: Design of experiments. *Neuroimage*. 2004; **21**:401–13.

[281] Liu H., Gao J. An investigation of the impulse functions for the nonlinear BOLD response in functional MRI. *Magn Reson Imaging*. 2000; **18**:931–8.

[282] Liu T. T., Frank L. R., Wong E. C., Buxton R. B. Detection power, estimation efficiency, and predictability in event-related fMRI. *Neuroimage*. 2001; **13**:759–73.

[283] Liu T. T., Frank L. R. Efficiency, power, and entropy in event-related fMRI with multiple trial types. Part I: Theory. *Neuroimage*. 2004; **21**:387–400.

[284] Logothetis N. K., Pauls J., Augath M., Trinath T., Oeltermann A. Neurophysiological investigation of the basis of the fMRI signal. *Nature* 2001; **412**:150–7.

[285] Logothetis N. K., Wandell B. A. Interpreting the BOLD signal. *Annu Rev Physiol*. 2004; **66**:735–69.

[286] Lohmann G., Bohn S. Using replicator dynamics for analyzing fMRI data of the human brain. *IEEE Trans Med Imag*. 2002; **21**:485–92.

[287] Lowe M. J., Sorenson J. A. Spatially filtering functional magnetic resonance imaging data. *Magn Reson Med*. 1997; **37**:723–9.

[288] Lowe M. J., Mock B. J., Sorenson J. A. Functional connectivity in single and multislice echoplanar imaging using resting-state fluctuations. *Neuroimage*. 1998; **7**:119–32.

[289] Lu Y., Zang Y., Jiang T. A modified temporal self-correlation method for analysis of fMRI time series. *Neuroinformatics*. 2003; **1**:259–69.

[290] Lukic A. S., Wernick M. N., Strother S. C. An evaluation of methods for detecting brain activations from functional neuroimages. *Artif Intell Med*. 2002; **25**:69–88.

[291] Lund T. E. fcMRI–mapping functional connectivity or correlating cardiac-induced noise? *Magn Reson Med*. 2001; **46**:628–9.

[292] Maas L. C., Renshaw P. F. Post-registration spatial filtering to reduce noise in functional MRI data sets. *Magn Reson Imaging*. 1999; **17**:1371–82.

[293] Macey P. M., Macey K. E., Kumar R., Harper R. M. A method for removal of global effects from fMRI time series. *Neuroimage*. 2004; **22**:360–6.

[294] Machulda M. M., Ward H. A., Cha R., O'Brien P., Jack C. R. Jr. Functional inferences vary with the method of analysis in fMRI. *Neuroimage*. 2001; **14**:1122–7.

[295] Maldjian J. A. Functional connectivity MR imaging: fact or artifact? *Am J Neuroradiol*. 2001; **22**:239–40.

[296] Mandeville J. B., Marota J. J., Ayata C., Zararchuk G., Moskowitz M. A., Rosen B., Weisskoff R. M. Evidence of a cerebrovascular postarteriole windkessel with delayed compliance. *J Cereb Blood Flow Metab*. 1999; **19**:679–89.

[297] Mansfield P. Multi-planar image formation using NMR spin echoes. *J Physics C*. 1977; **10**:L55.

[298] Marrelec G., Benali H., Ciuciu P., Pelegrini-Issac M., Poline J. B. Robust Bayesian estimation of the hemodynamic response function in event-related BOLD fMRI using basic physiological information. *Hum Brain Mapp*. 2003; **19**:1–17.

[299] Marrelec G., Ciuciu P., Pelegrini-Issac M., Benali H. Estimation of the hemodynamic response function in event-related functional MRI: directed acyclic graphs for a general Bayesian inference framework. *Inf Process Med Imaging*. 2003; **18**: 635–46.

[300] Marrelec G., Ciuciu P., Pelegrini-Issac M., Benali H. Estimation of the Hemodynamic Response Function in event-related functional MRI: Bayesian networks as a framework for efficient Bayesian modeling and inference. *IEEE Trans Med Imag*. 2004; **23**:959–67.

[301] McGonigle D. J., Howseman A. M., Athwal B. S., Friston K. J., Frackowiak R. S., Holmes A. P. Variability in fMRI: an examination of intersession differences. *Neuroimage*. 2000; **11**:708–34.

[302] McIntosh A. R., Bookstein F. L., Haxby J. V., Grady C. L. Spatial pattern analysis of functional brain images using partial least squares. *Neuroimage*. 1996; **3**:143–57.

[303] McKeown M. J., Makeig S., Brown G. G., Jung T. P., Kindermann S. S., Bell A. J., Sejnowski T. J. Analysis of fMRI data by blind separation into independent spatial components. *Hum Brain Mapp*. 1998; **6**:160–88.

[304] McKeown M. J., Sejnowski T. J. Independent component analysis of fMRI data: examining the assumptions. *Hum Brain Mapp*. 1998; **6**:368–72.

[305] McKeown M. J. Detection of consistently task-related activations in fMRI data with hybrid independent component analysis. *Neuroimage*. 2000; **11**:24–35.

[306] McKeown M. J., Varadarajan V., Huettel S., McCarthy G. Deterministic and stochastic features of fMRI data: implications for analysis of event-related experiments. *J Neurosci Meth*. 2002; **118**:103–13.

[307] McNamee R. L., Lazar N. A. Assessing the sensitivity of fMRI group maps. *Neuroimage*. 2004; **22**:920–31.

[308] Mechelli A., Price C. J., Friston K. J. Nonlinear coupling between evoked rCBF and BOLD signals: a simulation study of hemodynamic responses. *Neuroimage*. 2001; **14**:862–72.

[309] Mechelli A., Penny W. D., Price C. J., Gitelman D. R., Friston K. J. Effective connectivity and intersubject variability: using a multisubject network to test differences and commonalities. *Neuroimage*. 2002; **17**:1459–69.

[310] Mechelli A., Price C. J., Henson R. N., Friston K. J. Estimating efficiency a priori: a comparison of blocked and randomized designs. *Neuroimage*. 2003; **18**:798–805.

[311] Mechelli A., Henson R. N., Price C. J., Friston K. J. Comparing event-related and epoch analysis in blocked design fMRI. *Neuroimage*. 2003; **18**:806–10.

[312] Mechelli A., Price C. J., Noppeney U., Friston K. J. A dynamic causal modeling study on category effects: bottom-up or top-down mediation? *J Cogn Neurosci*. 2003; **15**:925–34.

[313] Menon R. S., Luknowski D. C., Gati J. S. Mental chronometry using latency-resolved functional MRI. *Proc Natl Acad Sci USA*. 1998; **95**:10902–7.

[314] Menon R. S., Kim S. G. Spatial and temporal limits in cognitive neuroimaging with fMRI. *Trends Cogn Sci*. 1999; **3**:207–216.

[315] Menon R. S., Goodyear B. G. Submillimeter functional localization in human striate cortex using BOLD contrast at 4 Tesla: implications for the vascular point-spread function. *Magn Reson Med*. 1999; **41**:230–5.

[316] Meyer F. G. Wavelet-based estimation of a semiparametric generalized linear model of fMRI time-series. *IEEE Trans Med Imag*. 2003; **22**:315–22.

[317] Meyer F. G., Chinrungrueng J. Analysis of event-related fMRI data using best clustering bases. *IEEE Trans Med Imag*. 2003; **22**:933–9.

[318] Miezin F. M., Maccotta L., Ollinger J. M., Petersen S. E., Buckner R. L. Characterizing the hemodynamic response: effects of presentation rate, sampling procedure,

and the possibility of ordering brain activity based on relative timing. *Neuroimage.* 2000; **11**:735–59.

[319] Mildner T., Norris D. G., Schwarzbauer C., Wiggins C. J. A qualitative test of the balloon model for BOLD-based MR signal changes at 3 T. *Magn Reson Med.* 2001; **46**:891–9.

[320] Miller K. L., Luh W. M., Liu T. T., Martinez A., Obata T., Wong E. C., Frank L. R., Buxton R. B. Nonlinear temporal dynamics of the cerebral blood flow response. *Hum Brain Mapp.* 2001; **13**:1–12.

[321] Mitra P. P., Pesaran B. Analysis of dynamic brain imaging data. *Biophys J.* 1999; **76**:691–708.

[322] Moller U., Ligges M., Georgiewa P., Grunling C., Kaiser W. A., Witte H., Blanz B. How to avoid spurious cluster validation? A methodological investigation on simulated and fMRI data. *Neuroimage.* 2002; **17**:431–46.

[323] Moriguchi H., Wendt M., Duerk J. L. Applying the uniform resampling (URS) algorithm to a Lissajous trajectory: fast image reconstruction with optimal gridding. *Magn Reson Med.* 2000; **44**:766–81.

[324] Moritz C. H., Haughton V. M., Cordes D., Quigley M., Meyerand M. E. Whole-brain functional MR imaging activation from a finger-tapping task examined with independent component analysis. *Am J Neuroradiol.* 2000; **21**:1629–35.

[325] Moritz C. H., Rogers B. P., Meyerand M. E. Power spectrum ranked independent component analysis of a periodic fMRI complex motor paradigm. *Hum Brain Mapp.* 2003; **18**:111–22.

[326] Morgan V. L., Pickens D. R., Hartmann S. L., Price R. R. Comparison of functional MRI image realignment tools using a computer-generated phantom. *Magn Reson Med.* 2001; **46**:510–4.

[327] Morgan V. L., Price R. R., Arain A., Modur P., Abou-Khalil B. Resting functional MRI with temporal clustering analysis for localization of epileptic activity without EEG. *Neuroimage.* 2004; **21**:473–81.

[328] Moser E., Diemling M., Baumgartner R. Fuzzy clustering of gradient-echo functional MRI in the human visual cortex. Part II: quantification. *J Magn Reson Imaging.* 1997; **7**:1102–8.

[329] Mudholkar G. S., George E. O. The logit method for combining probabilities. In *Symposium on Optimizing Methods in Statistics.* Rustagi J., Ed, New York, Academic Press. 1979; pp. 345–66.

[330] Muller H. P., Kraft E., Ludolph A., Erne S. N. New methods in fMRI analysis. Hierarchical cluster analysis for improved signal-to-noise ratio compared to standard techniques. *IEEE Eng Med Biol Mag.* 2002; **21**(5):134–42.

[331] Müller K., Lohmann G., Bosch V., von Cramon D. Y. On multivariate spectral analysis of fMRI time series. *Neuroimage.* 2001; **14**:347–56.

[332] Müller K., Lohmann G., Zysset S., von Cramon D. Y. Wavelet statistics of functional MRI data and the general linear model. *J Magn Reson Imaging.* 2003; **17**:20–30.

[333] Müller K., Lohmann G., Neumann J., Grigutsch M., Mildner T., von Cramon D. Y. Investigating the wavelet coherence phase of the BOLD signal. *J Magn Reson Imaging.* 2004; **20**:145–52.

[334] Munger P., Crelier G. R., Peters T. M., Pike G. B. An inverse problem approach to the correction of distortion in EPI images. *IEEE Trans Med Imag.* 2000; **19**:681–9.

[335] Nan F. Y., Nowak R. D. Generalized likelihood ratio detection for fMRI using complex data. *IEEE Trans Med Imag.* 1999; **18**:320–9.

[336] Nandy R. R., Cordes D. Novel ROC-type method for testing the efficiency of multivariate statistical methods in fMRI. *Magn Reson Med.* 2003; **49**:1152–62.

[337] Nestares O., Heeger D. J. Robust multiresolution alignment of MRI brain volumes. *Magn Reson Med*. 2000; **43**:705–15.

[338] Neumann J., Lohmann G. Bayesian second-level analysis of functional magnetic resonance images. *Neuroimage*. 2003; **20**:1346–55.

[339] Ngan S. C., Hu X. Analysis of functional magnetic resonance imaging data using self-organizing mapping with spatial connectivity. *Magn Reson Med*. 1999; **41**: 939–46.

[340] Ngan S. C., LaConte S. M., Hu X. Temporal filtering of event-related fMRI data using cross-validation. *Neuroimage*. 2000; **11**:797–804.

[341] Nichols T. E., Holmes A. P. Nonparametric permutation tests for functional neuroimaging: a primer with examples. *Hum Brain Mapp*. 2002; **15**:1–25.

[342] Nieto-Castanon A., Ghosh S. S., Tourville J. A., Guenther F. H. Region of interest based analysis of functional imaging data. *Neuroimage*. 2003; **19**:1303–16.

[343] Noll D. C., Genovese C. R., Nystrom L. E., Vazquez A. L., Forman S. D., Eddy W. F., Cohen J. D. Estimating test–retest reliability in functional MR imaging. II: Application to motor and cognitive activation studies. *Magn Reson Med*. 1997; **38**:508–17.

[344] Noll D. C., Peltier S. J., Boada F. E. Simultaneous multislice acquisition using rosette trajectories (SMART): a new imaging method for functional MRI. *Magn Reson Med*. 1998; **39**:709–16.

[345] Nowak R. D. Wavelet-based Rician noise removal for magnetic resonance imaging. *IEEE Trans Image Process*. 1999; **8**:1408–19.

[346] Obata T., Liu T. T., Miller K. L., Luh W. M., Wong E. C., Frank L. R., Buxton R. B. Discrepancies between BOLD and flow dynamics in primary and supplementary motor areas: application of the balloon model to the interpretation of BOLD transients. *Neuroimage*. 2004; **21**:144–53.

[347] Ogawa S., Lee T. M., Kay A. R., Tank D. W. Brain magnetic resonance imaging with contrast dependent on blood oxygenation. *Proc Natl Acad Sci USA*. 1990; **87**:9868–72.

[348] Ogawa S., Tank D. W., Menon R., Ellermann J. M., Kim S. G., Merkle H., Ugurbil K. Intrinsic signal changes accompanying sensory stimulation: functional brain mapping with magnetic resonance imaging. *Proc Natl Acad Sci USA*. 1992; **89**:5951–5.

[349] Ogawa S., Menon R. S., Tank D. W., Kim S. G., Merkle H., Ellermann J. M., Ugurbil K. Functional brain mapping by blood oxygenation level-dependent contrast magnetic resonance imaging. A comparison of signal characteristics with a biophysical model. *Biophys J*. 1993; **64**:803–12.

[350] Ogawa S., Menon R. S., Kim S. G., Ugurbil K. On the characteristics of functional magnetic resonance imaging of the brain. *Annu Rev Biophys Biomol Struct*. 1998; **27**:447–74.

[351] Ollinger J. M., Corbetta M., Shulman G. L. Separating processes within a trial in event-related functional MRI. *Neuroimage*. 2001; **13**:218–29.

[352] Ostuni J. L., Levin R. L., Frank J. A., DeCarli C. Correspondence of closest gradient voxels – a robust registration algorithm. *J Magn Reson Imaging*. 1997; **7**:410–5.

[353] Parker G. J., Wheeler-Kingshott C. A., Barker G. J. Estimating distributed anatomical connectivity using fast marching methods and diffusion tensor imaging. *IEEE Trans Med Imag*. 2002; **21**:505–12.

[354] Parrish T. B., Gitelman D. R., LaBar K. S., Mesulam M. M. Impact of signal-to-noise on functional MRI. *Mag Reson Med*. 2000; **44**:925–32.

[355] Pauling L., Coryell C. D. The magnetic properties and structure of the hemochromogens and related substances. *Proc Nat Acad Sci*. 1936; **22**:159–63.

[356] Penny W., Friston K. Mixtures of general linear models for functional neuroimaging. *IEEE Trans Med Imag*. 2003; **22**:504–14.

[357] Penny W., Kiebel S., Friston K. Variational Bayesian inference for fMRI time series. *Neuroimage*. 2003; **19**:727–41.

[358] Penny W. D., Stephan K. E., Mechelli A., Friston K. J. Comparing dynamic causal models. *Neuroimage*. 2004; **22**:1157–72.

[359] Petersson K. M., Nichols T. E., Poline J. B., Holmes A. P. Statistical limitations in functional neuroimaging. I. Non-inferential methods and statistical models. *Phil Trans R Soc Lond B Biol Sci*. 1999; **354**:1239–60.

[360] Pfeuffer J., McCullough J. C., Van de Moortele P. F., Ugurbil K., Hu X. Spatial dependence of the nonlinear BOLD response at short stimulus duration. *Neuroimage*. 2003; **18**:990–1000.

[361] Poline J. B., Worsley K. J., Holmes A. P., Frackwiak R. S., Friston K. J. Estimating smoothness in statistical parametric maps: variability of *p*-values. *J Comput Tomogr*. 1995; **19**:788–96.

[362] Poline J. B., Worsley K. J., Evans A. C., Friston K. J. Combining spatial extent and peak intensity to test for activations in functional imaging. *Neuroimage*. 1997; **5**:83–96.

[363] Press W. H., Flannery B. P., Teukolsky S. A., Vetterling W. T. *Numerical Recipes in C: The Art of Scientific Computing*. Cambridge, Cambridge University Press. 1992.

[364] Price C. J., Veltman D. J., Ashburner J., Josephs O., Friston K. J. The critical relationship between the timing of stimulus presentation and data acquisition in blocked designs with fMRI. *Neuroimage*. 1999; **10**:36–44.

[365] Purdon P. L., Weisskoff R. M. Effect of temporal autocorrelation due to physiological noise and stimulus paradigm on voxel-level false-positive rates in fMRI. *Hum Brain Mapp*. 1998; **6**:239–49.

[366] Purdon P. L., Solo V., Weisskoff R. M., Brown E. N. Locally regularized spatio-temporal modeling and model comparison for functional MRI. *Neuroimage*. 2001; **14**:912–23.

[367] Rajapakse J. C., Piyaratna J. Bayesian approach to segmentation of statistical parametric maps. *IEEE Trans Biomed Eng*. 2001; **48**:1186–94.

[368] Ramsay J. O., Silverman B. W. *Functional Data Analysis*. New York, Springer-Verlag. 1997.

[369] Reber P. J., Wong E. C., Buxton R. B., Frank L. R. Correction of off resonance-related distortion in echo-planar imaging using EPI-based field maps. *Magn Reson Med*. 1998; **39**:328–30.

[370] Reber P. J., Wong E. C., Buxton R. B. Encoding activity in the medial temporal lobe examined with anatomically constrained fMRI analysis. *Hippocampus*. 2002; **12**:363–76.

[371] Rencher A. C. *Multivariate Statistical Inference and Applications*. New York, John-Wiley and Sons. 1998.

[372] Rencher A. C. *Methods of Multivariate Analysis*, 2nd Edition. New York, John-Wiley and Sons. 2002.

[373] Richter W., Richter M. The shape of the fMRI BOLD response in children and adults changes systematically with age. *Neuroimage*. 2003; **20**:1122–31.

[374] Riera J. J., Watanabe J., Kazuki I., Naoki M., Aubert E., Ozaki T., Kawashima R. A state-space model of the hemodynamic approach: nonlinear filtering of BOLD signals. *Neuroimage*. 2004; **21**:547–67.

[375] Roland P. E., Graufelds C. J., Wahlin J., Ingelman L., Andersson M., Ledberg A., Pedersen J., Akerman S., Dabringhaus A., Zilles K. Human brain atlas: for

high resolution functional and anatomical mapping. *Hum Brain Mapp.* 1994;
1:173–84.

[376] Rosen B. R., Buckner R. L., Dale A. M. Event-related functional MRI: past, present, and future. *Proc Natl Acad Sci USA.* 1998; **95**:773–80.

[377] Rowe D. B. Bayesian source separation for reference function determination in fMRI. *Magn Reson Med.* 2001; **46**:374–8.

[378] Ruan S., Jaggi C., Constans J. M., Bloyet D. Detection of brain activation signal from functional magnetic resonance imaging data. *J Neuroimaging.* 1996; **6**:207–12.

[379] Rudin W. *Real and Complex Analysis.* New York, McGraw-Hill. 1987.

[380] Ruttimann U. E., Unser M., Rawlings R. R., Rio D., Ramsey N. F., Mattay V. S., Hommer D. W., Frank J. A., Weinberger D. R. Statistical analysis of functional MRI data in the wavelet domain. *IEEE Trans Med Imag.* 1998; **17**:142–54.

[381] Saad Z. S., Ropella K. M., Cox R. W., DeYoe E. A. Analysis and use of FMRI response delays. *Hum Brain Mapp.* 2001; **13**:74–93.

[382] Saad Z. S., Ropella K. M., DeYoe E. A., Bandettini P. A. The spatial extent of the BOLD response. *Neuroimage.* 2003; **19**:132–44.

[383] Salli E., Aronen H. J., Savolainen S., Korvenoja A., Visa A. Contextual clustering for analysis of functional MRI data. *IEEE Trans Med Imag.* 2001; **20**:403–14.

[384] Salli E., Korvenoja A., Visa A., Katila T., Aronen H. J. Reproducibility of fMRI: effect of the use of contextual information. *Neuroimage.* 2001; **13**:459–71.

[385] Sarty G. E. Reconstruction of nuclear magnetic resonance imaging data from non-Cartesian grids. In *Advances in Imaging and Electron Physics*, Vol. III. Hawkes P., Ed, New York, Academic Press. 1999; pp. 231–345.

[386] Sarty G. E. Critical sampling in ROSE scanning. *Magn Reson Med.* 2000; **44**:129–36.

[387] Sarty G. E. The point spread function of convolution regridding reconstruction. In *Signal Processing for Magnetic Resonance Imaging and Spectroscopy.* Yan H., Ed, New York, Marcel Dekker. 2002; pp. 59–89.

[388] Sarty G. E. Single TrAjectory Radial (STAR) imaging. *Magn Reson Med.* 2004; **51**:445–51.

[389] Sarty G. E., Borowsky R. Functional MRI activation maps from empirically defined curve fitting. *Concepts Mag Reson Part B (Magnetic Resonance Engineering).* 2005; **24B**:46–55.

[390] Schacter D. L., Buckner R. L., Koutstaal W. Memory, consciousness and neuroimaging. *Phil Trans R Soc Lond B Biol Sci.* 1998; **353**:1861–78.

[391] Schetzen M. *The Volterra and Wiener Theories of Nonlinear Systems.* New York, Wiley. 1980.

[392] Schlosser R., Gesierich T., Kaufmann B., Vucurevic G., Hunsche S., Gawehn J., Stoeter P. Altered effective connectivity during working memory performance in schizophrenia: a study with fMRI and structural equation modeling. *Neuroimage.* 2003; **19**:751–63.

[393] Schmithorst V. J., Dardzinski B. J., Holland S. K. Simultaneous correction of ghost and geometric distortion artifacts in EPI using a multiecho reference scan. *IEEE Trans Med Imag.* 2001; **20**:535–9.

[394] Schwarzbauer C., Heinke W. Investigating the dependence of BOLD contrast on oxidative metabolism. *Magn Reson Med.* 1999; **41**:537–43.

[395] Segebarth C., Belle V., Delon C., Massarelli R., Decety J., Le Bas J. F., Decorps M., Benabid A. L. Functional MRI of the human brain: predominance of signals from extracerebral veins. *Neuroreport.* 1994; **5**:813–6.

[396] Serences J. T. A comparison of methods for characterizing the event-related BOLD time series in rapid fMRI. *Neuroimage.* 2004; **21**:1690–1700.

[397] Shimizu Y., Barth M., Windischberger C., Moser E., Thurner S. Wavelet-based multifractal analysis of fMRI time series. *Neuroimage*. 2004; **22**:1195–202.

[398] Shimony J. S., Snyder A. Z., Conturo T. E., Corbetta M. The study of neural connectivity using diffusion tensor tracking. *Cortex*. 2004; **40**:213–5.

[399] Shmuel A., Yacoub E., Pfeuffer J., Van de Moortele P. F., Adriany G., Hu X., Ugurbil K. Sustained negative BOLD, blood flow and oxygen consumption response and its coupling to the positive response in the human brain. *Neuron*. 2002; **36**: 1195–210.

[400] Silva A. C., Barbier E. L., Lowe I. J., Kortesky A. P. Radial echo-planar imaging. *J Magn Reson* 1998; **135**:242–7.

[401] Skudlarski P., Constable R. T., Gore J. C. ROC analysis of statistical methods used in functional MRI: individual subjects. *Neuroimage*. 1999; **9**:311–29.

[402] Smith A. M., Lewis B. K., Rittimann U. E., Ye F. Q., Sinnwell T. M., Yang Y., Duyn J. H., Frank J. A. Investigation of low frequency drift in fMRI signal. *Neuroimage* 1999; **9**:526–33.

[403] Smith M., Putz B., Auer D., Fahrmeir L. Assessing brain activity through spatial Bayesian variable selection. *Neuroimage*. 2003; **20**:802–15.

[404] Smyser C., Grabowski T. J., Frank R. J., Haller J. W., Bolinger L. Real-time multiple linear regression for fMRI supported by time-aware acquisition and processing. *Magn Reson Med*. 2001; **45**:289–98.

[405] Solé A. F., Ngan S. C., Sapiro G., Hu X., Lopez A. Anisotropic 2-D and 3-D averaging of fMRI signals. *IEEE Trans Med Imag*. 2001; **20**:86–93.

[406] Solo V., Purdon P., Weisskoff R., Brown E. A signal estimation approach to functional MRI. *IEEE Trans Med Imag*. 2001; **20**:26–35.

[407] Soltysik D. A., Peck K. K., White K. D., Crosson B., Briggs R. W. Comparison of hemodynamic response non-linearity across primary cortical areas. *Neuroimage*. 2004; **22**:1117–27.

[408] Stanberry L., Nandy R., Cordes D. Cluster analysis of fMRI data using dendrogram sharpening. *Hum Brain Mapp*. 2003; **20**:201–19.

[409] Stenger V. A., Peltier S., Boada F. E., Noll D. C. 3D spiral cardiac/respiratory ordered fMRI data acquisition at 3 Tesla. *Magn Reson Med*. 1999; **41**:983–91.

[410] Stouffer S. A., Suchman E. A., DeVinney L. C., Star S. A., Williams R. M. *The American Soldier*: Vol. I. *Adjustment During Army Life*. Princeton, Princeton University Press. 1949.

[411] Strupp J. P. Stimulate: A GUI based fMRI analysis software package. *Neuroimage*. 1996; **3**:S607.

[412] Studholme C., Constable R. T., Duncan J. S. Accurate alignment of functional EPI data to anatomical MRI using a physics-based distortion model. *IEEE Trans Med Imag*. 2000; **19**:1115–27.

[413] Suckling J., Bullmore E. Permutation tests for factorially designed neuroimaging experiments. *Hum Brain Mapp*. 2004; **22**:193–205.

[414] Sun F. T., Miller L. M., D'Esposito M. Measuring interregional functional connectivity using coherence and partial coherence analyses of fMRI data. *Neuroimage*. 2004; **21**:647–58.

[415] Svensen M., Kruggel F., Benali H. ICA of fMRI group study data. *Neuroimage*. 2002; **16**:551–63.

[416] Talairach J., Tournoux P. *Co-Planar Stereotaxic Atlas of the Human Brain*. New York, Thieme Medical Publishers. 1988.

[417] Tanabe J., Miller D., Tregellas J., Freedman R., Meyer F. G. Comparison of detrending methods for optimal fMRI preprocessing. *Neuroimage*. 2002; **15**:902–7.

[418] Taylor J. G., Krause B., Shah N. J., Horwitz B., Mueller-Gaertner H. W. On the relation between brain images and brain neural networks. *Hum Brain Mapp*. 2000; **9**:165–82.

[419] Tegeler C., Strother S. C., Anderson J. R., Kim S. G. Reproducibility of BOLD-based functional MRI obtained at 4 T. *Hum Brain Mapp*. 1999; **7**:267–83.

[420] Thevenaz P., Ruttimann U. E., Unser M. A pyramid approach to subpixel registration based on intensity. *IEEE Trans Image Proc*. 1998; **7**:27–41.

[421] Thomas C. G., Harshman R. A., Menon R. S. Noise reduction in BOLD-based fMRI using component analysis. *Neuroimage*. 2002; **17**:1521–37.

[422] Tippett L. H. C. The Method of Statistics. London, Williams and Norgate. 1931.

[423] Toronov V., Walker S., Gupta R., Choi J. H., Gratton E., Hueber D., Webb A. The roles of changes in deoxyhemoglobin concentration and regional cerebral blood volume in the fMRI BOLD signal. *Neuroimage*. 2003; **19**:1521–31.

[424] von Tscharner V., Thulborn K. R. Specified-resolution wavelet analysis of activation patterns from BOLD contrast fMRI. *IEEE Trans Med Imag*. 2001; **20**:704–14.

[425] Tzourio-Mazoyer N., Landeau B., Papathanassiou D., Crivello F., Etard O., Delcroix N., Mazoyer B., Joliot M. Automated anatomical labeling of activations in SPM using a macroscopic anatomical parcellation of the MNI MRI single-subject brain. *Neuroimage*. 2002; **15**:273–89.

[426] Ugurbil K., Ogawa S., Kim S.-G., Hu X., Chen W., Zhu X.-X. Imaging brain activity using nuclear spins. In: *Magnetic Resonance and Brain Function: Approaches from Physics*. Maraviglia B. Ed. Amsterdam, BIOS Press. 1999; pp. 261–310.

[427] Uludağ K., Dubowitz D. J., Yoder E. J., Restom K., Liu T. T., Buxton R. B. Coupling of cerebral blood flow and oxygen consumption during physiological activation and deactivation measured with fMRI. *Neuroimage*. 2004; **23**:148–55.

[428] Valdes-Sosa P. A. Spatio-temporal autoregressive models defined over brain manifolds. *Neuroinformatics*. 2004; **2**:239–50.

[429] Vazquez A. L., Noll D. C. Nonlinear aspects of the BOLD response in functional MRI. *Neuroimage*. 1998; **7**:108–18.

[430] Veltman D. J., Mechelli A., Friston K. J., Price C. J. The importance of distributed sampling in blocked functional magnetic resonance imaging designs. *Neuroimage*. 2002; **17**:1203–6.

[431] van de Ven V. G., Formisano E., Prvulovic D., Roeder C. H., Linden D. E. Functional connectivity as revealed by spatial independent component analysis of fMRI measurements during rest. *Hum Brain Mapp*. 2004; **22**:165–78.

[432] Visscher K. M., Miezin F. M., Kelly J. E., Buckner R. L., Donaldson D. I., McAvoy M. P., Bhalodia V. M., Petersen S. E. Mixed blocked/event-related designs separate transient and sustained activity in fMRI. *Neuroimage*. 2003; **19**: 1694–708.

[433] Vlaardingerbroek M. T., den Boer J. A. *Magnetic Resonance Imaging: Theory and Practice*. Berlin, Springer. 2003.

[434] Voyvodic J. T. Real-time fMRI paradigm control, physiology, and behavior combined with near real-time statistical analysis. *Neuroimage*. 1999; **10**:91–106.

[435] Wade A. R. The negative BOLD signal unmasked. *Neuron*. 2002; **36**:993–5.

[436] Wager T. D., Nichols T. E. Optimization of experimental design in fMRI: a general framework using a genetic algorithm. *Neuroimage*. 2003; **18**:293–309.

[437] Weaver J. B. Monotonic noise suppression used to improve the sensitivity of fMRI activation maps. *J Digit Imag*. 1998; **11**:46–52.

[438] Weaver J. B., Xu Y., Healy D. M., Cromwell L. D. Filtering noise from images with wavelet transforms. *Magn Reson Med*. 1991; **21**:288–95.

[439] Wenger K. K., Visscher K. M., Miezin F. M., Petersen S. E., Schlaggar B. L. Comparison of sustained and transient activity in children and adults using a mixed blocked/event-related fMRI design. *Neuroimage.* 2004; **22**:975–85.

[440] West M., Harrison J. *Bayesian Forecasting and Dynamic Models.* Heidelberg, Springer-Verlag. 1997.

[441] Westin C. F., Maier S. E., Mamata H., Nabavi A., Jolesz F. A., Kikinis R. Processing and visualization for diffusion tensor MRI. *Med Image Anal.* 2002; **6**:93–108.

[442] Wilkinson B. A statistical consideration in psychological research. *Psychol Bull.* 1951; **48**:156–8.

[443] Windischberger C., Barth M., Lamm C., Schroeder L., Bauer H., Gur R. C., Moser E. Fuzzy cluster analysis of high-field functional MRI data. *Artif Intell Med.* 2003; **29**:203–23.

[444] Wink A. M., Roerdink J. B. Denoising functional MR images: a comparison of wavelet denoising and Gaussian smoothing. *IEEE Trans Med Imag.* 2004; **23**:374–87.

[445] Wise R. G., Ide K., Poulin M. J., Tracey I. Resting fluctuations in arterial carbon dioxide induce significant low frequency variations in BOLD signal. *Neuroimage.* 2004; **21**:1652–64.

[446] Wismüller A., Meyer-Base A., Lange O., Auer D., Reiser M. F., Sumners D. Model-free functional MRI analysis based on unsupervised clustering. *J Biomed Inform.* 2004; **37**:10–8.

[447] Woods R. P., Grafton S. T., Holmes C. J., Cherry S. R., Mazziotta J. C. Automated image registration: I. General methods and intrasubject, intramodality validation. *J Comput Assist Tomogr.* 1998; **22**:139–52.

[448] Woods R. P., Grafton S. T., Watson J. D. G., Sicotte N. L., Mazziotta J. C. Automated image registration: II. Intersubject validation of linear and non-linear models. *J Comput Asst Tomogr.* 1998; **22**:153–65.

[449] Woolrich M. W., Ripley B. D., Brady M., Smith S. M. Temporal autocorrelation in univariate linear modeling of FMRI data. *Neuroimage.* 2001; **14**:1370–86.

[450] Woolrich M. W., Behrens T. E., Beckmann C. F., Jenkinson M., Smith S. M. Multilevel linear modeling for FMRI group analysis using Bayesian inference. *Neuroimage.* 2004; **21**:1732–47.

[451] Woolrich M. W., Behrens T. E., Smith S. M. Constrained linear basis sets for HRF modeling using Variational Bayes. *Neuroimage.* 2004; **21**:1748–61.

[452] Woolrich M. W., Jenkinson M., Brady J. M., Smith S. M. Fully Bayesian spatiotemporal modeling of FMRI data. *IEEE Trans Med Imag.* 2004; **23**:213–31.

[453] Worsley K. J. Detecting activation in fMRI data. *Stat Methods Med Res.* 2003; **12**:401–18.

[454] Worsley K. J., Evans A. C., Marrett S., Neelin P. A three-dimensional statistical analysis for CBF activation studies in human brain. *J Cereb Blood Flow Metab.* 1992; **12**:900–18.

[455] Worsley K. J., Friston K. J. Analysis of fMRI time-series revisited–again. *Neuroimage.* 1995; **2**:173–81.

[456] Worsley K. J., Cao J., Paus T., Petrides M., Evans A. C. Applications of random field theory to functional connectivity. *Hum Brain Mapp.* 1998; **6**:364–7.

[457] Worsley K. J., Andermann M., Koulis T., MacDonald D., Evans A. C. Detecting changes in nonisotropic images. *Hum Brain Mapp.* 1999; **8**:98–101.

[458] Worsley K. J., Friston K. J. A test for conjunction. *Stat Probability Lett.* 2000; **47**:135–40.

[459] Worsley K. J., Liao C. H., Aston J., Petre V., Duncan G. H., Morales F., Evans A. C. A general statistical analysis for fMRI data. *Neuroimage.* 2002; **15**:1–15.

[460] Wowk B., McIntyre M. C., Saunders J. K. *K*-space detection and correction of physiological artifacts in fMRI. *Magn Reson Med.* 1997; **38**:1029–34.

[461] Wu G., Luo F., Li Z., Zhao X., Li S. J. Transient relationships among BOLD, CBV, and CBF changes in rat brain as detected by functional MRI. *Magn Reson Med.* 2002; **48**:987–93.

[462] Xiong J., Parsons L. M., Gao J. H., Fox P. T. Interregional connectivity to primary motor cortex revealed using MRI resting state images. *Hum Brain Mapp.* 1999; **8**:151–6.

[463] Xiong J., Fox P. T., Gao J. H. Directly mapping magnetic field effects of neuronal activity by magnetic resonance imaging. *Hum Brain Mapp.* 2003; **20**:41–49.

[464] Yee S. H., Gao J. H. Improved detection of time windows of brain responses in fMRI using modified temporal clustering analysis. *Magn Reson Imaging.* 2002; **20**:17–26.

[465] Yetkin F. Z., McAuliffe T. L., Cox R., Haughton V. M. Test–retest precision of functional MR in sensory and motor task activation. *Am J Neuroradiol.* 1996; **17**:95–8.

[466] Yetkin F. Z., Haughton V. M., Cox R. W., Hyde J., Birn R. M., Wong E. C., Prost R. Effect of motion outside the field of view on functional MR. *Am J Neuroradiol.* 1996; **17**:1005–9.

[467] Yoo S. S., Guttmann C. R., Panych L. P. Multiresolution data acquisition and detection in functional MRI. *Neuroimage.* 2001; **14**:1476–85.

[468] Zang Y., Jiang T., Lu Y., He Y., Tian L. Regional homogeneity approach to fMRI data analysis. *Neuroimage.* 2004; **22**:394–400.

[469] Zarahn E., Aguirre G. K., D'Esposito M. Empirical analyses of BOLD fMRI statistics. I. Spatially unsmoothed data collected under null-hypothesis conditions. *Neuroimage.* 1997; **5**:179–197.

[470] Zaroubi S., Goelman G. Complex denoising of MR data via wavelet analysis: application for functional MRI. *Magn Reson Imaging.* 2000; **18**:59–68.

[471] Zeng H, Constable RT. Image distortion correction in EPI: comparison of field mapping with point spread function mapping. *Magn Reson Med.* 2002; **48**:137–46.

[472] Zhang R., Cox R. W., Hyde J. S. The effect of magnetization transfer on functional MRI signals. *Magn Reson Med.* 1997; **38**:187–92.

[473] Zhao X., Glahn D., Tan L. H., Li N., Xiong J., Gao J. H. Comparison of TCA and ICA techniques in fMRI data processing. *J Magn Reson Imaging.* 2004; **19**:397–402.

[474] Zheng Y., Martindale J., Johnston D., Jones M., Berwick J., Mayhew J. A model of the hemodynamic response and oxygen delivery to brain. *Neuroimage.* 2002; **16**:617–37.

[475] Zhilkin P., Alexander M. E. 3D image registration using a fast noniterative algorithm. *Magn Reson Imaging.* 2000; **18**:1143–50.

[476] Zhilkin P., Alexander M. E. Affine registration: a comparison of several programs. *Magn Reson Imaging.* 2004; **22**:55–66.

Index

185